Key Challenges in Geography

EUROGEO Book Series

Series Editors

This book series addresses relevant topics in the wide field of geography, which connects the physical, human and technological sciences to enhance teaching, research, and decision making. Geography provides answers to how aspects of these sciences are interconnected and are forming spatial patterns and processes that have impact on global, regional and local issues and thus affect present and future generations. Moreover, by dealing with places, people and cultures, Geography explores international issues ranging from physical, urban and rural environments and their evolution, to climate, pollution, development and political economy.

Key Challenges in Geography is an initiative of the European Association of Geographers (EUROGEO), an organization dealing with examining geographical issues from a European perspective, representing European Geographers working in different professional activities and at all levels of education. EUROGEO's goal and the core part of its statutory activities is to make European Geography a worldwide reference and standard. The book series serves as a platform for members of EUROGEO as well as affiliated National Geographical Associations in Europe, but is equally open to contributions from non-members.

The book series addresses topics of contemporary relevance in the wide field of geography. It has a global scope and includes contributions from a wide range of theoretical and applied geographical disciplines.

Key Challenges in Geography aims to:

- present collections of chapters on topics that reflect the significance of Geography as a discipline;
- provide disciplinary and interdisciplinary titles related to geographical, environmental, cultural, economic, political, urban and technological research with a European dimension, but not exclusive;
- deliver thought-provoking contributions related to cross-disciplinary approaches and interconnected works that explore the complex interactions among geography, technology, politics, environment and human conditions;
- publish volumes tackling urgent topics to geographers and policy makers alike;
- publish comprehensive monographs, edited volumes and textbooks refereed by European and worldwide experts specialized in the subjects and themes of the books;
- provide a forum for geographers worldwide to communicate on all aspects of research and applications of geography, with a European dimension, but not exclusive.

All books/chapters will undergo a blind review process with a minimum of two reviewers.

An author/editor questionnaire, instructions for authors and a book proposal form can be obtained by contacting the Publisher.

More information about this series at http://www.springer.com/series/15694

Antonio Augusto Rossotto Ioris

Frontier Making in the Amazon

Economic, Political and Socioecological Conversion

EUROGEO

Springer

Antonio Augusto Rossotto Ioris
School of Geography and Planning
Cardiff University
Cardiff, UK

ISSN 2522-8420 ISSN 2522-8439 (electronic)
Key Challenges in Geography
ISBN 978-3-030-38523-1 ISBN 978-3-030-38524-8 (eBook)
https://doi.org/10.1007/978-3-030-38524-8

This Springer imprint is published by the registered company Springer Nature Switzerland AG
The registered company address is: Gewerbestrasse 11, 6330 Cham, Switzerland

Contents

Abbreviations

BR	Code of the Brazilian federal motorways
CIMI	Indigenist Missionary Council
CPE	Cultural political economy
FUNAI	National Indian Foundation
GDP	Gross domestic product
GMO	Genetically modified organism
ha	Hectares
IBAMA	Brazilian Institute of the Environment and Renewable Natural Resources
LSA	'Law of scarcity–abundance'
MP	Provisional Decree (introduced by the president and later ratified/rejected by the National Congress)
PAC	Growth Acceleration Programme
PFF	Peasant family farming
RESEX	Extractive reserve
SEMA	Mato Grosso's environmental protection agency
TNCs	Transnational corporations
UFMT	Federal University of Mato Grosso, Brazil
UPRB	Upper Paraguay River Basin

List of Figures

List of Tables

Chapter 1
Introduction: Frontier Thinking and the Amazon Region

Abstract The chapter introduces the aims and content of the book, highlighting its focus on peripheral socio-economic areas and marginal regions where development, as traditionally defined, is still in the nascent stages. These peripheries constitute new frontiers for economic development and typically unveil a disconcerting combination of change and continuity, if not frustration. Frontiers contain the germ of renovation, the chance to begin something anew and avoid the failures experienced elsewhere. However, almost invariably, the result is a missed opportunity to enact a different trajectory, and the frontier becomes a spectre of what it could have been. Because socio-spatial frontiers emerge, primarily, to mitigate and ameliorate the troubles and insufficiencies that characterise the politico-economic centres from where people, capital and institutions originated, the central claim of the book is that frontiers are characterised by built-in obsolescence and fleeting hopes of renewal. This is the most essential paradox, among many others, of the genesis of frontiers: the frontier space is pregnant with opportunities and possibilities, but in most cases, it delivers only stillborn dreams and expectations.

Keywords Political geography · Political ecology · Colonialism · Globalisation · Territorialisation

This book is an investigation into peripheral socio-economic areas and marginal regions where development, as traditionally defined, is still in the nascent stages. These peripheries constitute new frontiers for economic development and typically unveil a disconcerting combination of change and continuity, if not frustration. The frontier is perceived as a space where things 'as we know them' have not happened, or fully happened, yet. It contains the germ of renovation, the chance to begin something anew and, with a little luck, avoid the failures experienced elsewhere. However, because socio-spatial frontiers emerge, primarily, to mitigate and ameliorate the troubles and insufficiencies that characterise the politico-economic centres from where people, capital and institutions originated, our central claim throughout this text is that frontiers, in a capitalist economy, are characterised by built-in obsolescence and fleeting hopes of renewal. This is the most essential paradox, among many others, of the genesis of frontiers: the frontier space is pregnant with opportunities and possibilities, but in most cases, it delivers only stillborn dreams and expectations. The

A. A. R. Ioris, *Frontier Making in the Amazon*, Key Challenges in Geography,
https://doi.org/10.1007/978-3-030-38524-8_1

future is announced, but the reality of the experience frustrates the majority of those involved, leaving any unfulfilled goals to be shifted to the next available frontier. What happens in the periphery can also be metaphorically associated with a child raised by a problematic family. It is very likely that her life will be significantly affected by the failings of her parents and the adults close to her, but at the same time, there is always some hope that the new existence could become an improved version of the previous generations. Almost invariably, the result is a missed opportunity to enact a different trajectory, and the frontier becomes a spectre of what it could have been.

The puzzling basis of development frontiers, this intriguing fusion of ongoing transformations and indefinite prospects, makes them an eminent priority for research and discussion. Although often poorly understood, frontiering (i.e. the materialisation of new frontiers; also known as frontier making) frequently attracts great curiosity, if not fascination, from academics, officeholders and general society. The public knows that each frontiering experience is unique, mediated and fought for; it is also an intricate open-ended process, rather than a well-defined historical period. In addition, the ontological basis of frontier making has heuristic features that reveal a great deal about the shortcomings of development and the difficulty of pursuing politico-economic alternatives. Frontiers are like the Aleph in Jorge Luis Borges' short story (1949), places that contain the simultaneous times and circumstances of other locations. From the Aleph, it is possible to see and even sense the multiplicity of developments and the accumulation of tensions around the world (Interestingly, after becoming overwhelmed with the experience, the narrator of the story pretends to have seen nothing and his life continues unchanged, which is not too different from society's regular inability to learn from the failures of frontier making.). The word frontier, just like progress or equality, has imprecise and debatable connotations (Tanca 2011), related to expressions such as 'the limits of a settled land beyond which lies wilderness' and 'the extreme edge of understanding or achievements in a particular discipline'. Etymologically, frontier is something that is 'in front', an area that is part of the whole that is ahead in the hinterland or in a foreign location. A frontier is outward-oriented (i.e. directed towards outlying areas that are both a source of danger and a desired prize), while a boundary is inward-oriented, created and maintained by states and governments, with no life of its own (Kristof 1959). Both frontiers and boundaries are sociopolitical relations, although frontiers are meant to integrate, while borders are instituted to separate. The term is also used to describe technological innovations, the expansion of digital services, moral or behavioural limits, migratory movements and international relations, among other connotations (e.g. Kaltoft 2001).

Beyond ontological debates, the quest for economic and social frontiers is something recurrent in different cultures and contexts. It certainly precedes the history of capitalism, although frontiering has been an essential component of the expansion and strengthening of capital. Old civilisations left evidence of regular movement by parts of the population to different regions and places where social life was expected to resume and flourish. New frontspaces (i.e. the spaces of the frontier) and the belief that they would remain available in the future were always considered, even implicitly, an economic, social and political lifeline for many societies. Among others, the history of the Jewish people, for more than three millennia, has been characterised by the continuous resurgence of socio-spatial frontiers. From the conquest by the Egyptians, the Assyrian captivity, the Babylonian exile and the Roman occupation to the struggle during the Middle Ages and their integration in the Modern world, the Jews have had to coexist with other peoples and try to reorganise their lives in different territories, while many also aspired to return to their ancient land. The frontier for the Jews was both the new locations that followed conquest and diaspora and also the dream of reconstructing their nation in the olden land of Palestine (a territory currently disputed by several antagonistic parties). The evolution of the Portuguese nation (mentioned in the next chapter) can likewise be associated with the constraints of a small mainland territory ruled by a retrograde politico-economic elite who at the same time championed the conquest and subjugation of economic frontiers on all continents. This national drama was immortalised in Luis de Camões' epic poem The Lusiads, which celebrates the navigations and discoveries by Portuguese explorers which helped to consolidate the European Renaissance and propel Lusitan cultural values.

Literary texts certainly contain precious examples of the pursuit of frontier places for the reconstruction of societies and revitalisation of relations. For instance, Guimarães Rosa's The Devil to Pay in the Backlands [*Grande Sertão: Veredas*] of 1956—a masterpiece frequently compared to James Joyce's Ulysses, which also deals with existential and cultural frontiers—is an interminable journey in search of a different spatial order and personal and collective truth. In the book, things, identities and language are in constant disarray, but everything comes together because of the collective movement of the group, exploring territorial, symbolic and moral frontiers. Similarly, Tolstoy, in Anna Karenina, makes reference to the remote parts of Russia where people were barely touched by capitalist relations, perceived as primitive and unsophisticated but representing a splendid reserve of workers available to be incorporated into the national modernisation process:

> Though he had to admit that in the greater part of Russia, the eastern part, income was still zero, that for nine-tenths of the Russian population of eighty million wages were only at subsistence level, and that capital did not exist otherwise than as the most primitive tools – he still regarded all workers from that point of view alone. (Tolstoy 2006: 679)

Both in the novels and in real life, frontier making involves a multiplicity of symbols and forces that are put into action to nourish a general sense of renovation. It is common to associate frontiers with images of heroism and progress overcoming the past and defeating an archaic and exotic pre-frontier situation. However, the hint of newness and the promises of improvement associated with frontiers are highly qualified. New frontiers continue to be constantly incorporated into the economy but basically to maintain established patterns of production, allocation and consumption. Despite the flavour of modernity, as found in narratives about the conquest of Australia, Africa and the Americas, frontspaces have replicated processes of violence, accumulation and political control that had historically dominated political and economic relations in central areas (Ioris and Ioris 2018). It is as if the modern was emphatically announced, but the old systematically reemerged. The socio-economic modernisation bestowed by an ever-expanding world market is based on the paradoxical 'unity of disunity', 'a maelstrom of perpetual disintegration and renewal, of struggle and contradiction, of ambiguity and anguish' (Anderson 1984: 97). The formidable expansion of globalisation continues to be largely based on socio-spatial frontiers through the subordination of the Global South to new rounds of primary products export, consumption of goods and the circulation of capital that perpetuate injustices and aggravate environmental degradation (Verma 2019), while in the central areas of Europe and North America, the working class has been demobilised and confined to social and spatial ghettos (Guilluy 2014). Emblematic examples of the maelstrom of globalised trends associated with frontier making today are the recolonisation of Iraq after the 2003 war, the deindustrialisation and reprimarisation of the national economies of Brazil and Russia and the expansion of labour exploitation in industrial areas of China and other Asian countries.

At this point, it is appropriate to briefly review some disciplinary handlings of frontier and frontier making, which expose the contrasting epistemological bases and differences in conceptualisation between various scholarly approaches. These can be schematically divided between broad historic-epical accounts, diplomatic–legalistic narratives, economic explanations and geographical interpretations. First, the historical accounts of the opening of spatial frontiers are typically associated with the advance of progress and development along the lines of Western economic pressures and growing demands for resources, as well as opportunities for migrants to settle and recreate livelihoods within the new socio-spatial context. Frontiers are a zone or territory of interpenetration between distinct societies and a blend of multiple agents, including rich and poor explorers (Dobson and Maxfield 2005; Guelke 1976; Jordan and Kaups 1988; Lamar and Thompson 1981; Lester 1998; Miller and Steffen 1977; Thorpe 1996; van Lier 1971). The classical example in this literature was the 'frontier thesis' of Frederick Turner (1920), which argued that American society not only relied on frontier making, but has been essentially shaped by the constant attempt to conquer more land, subjugate local peoples and make space for economic adventurers. According to Turner, the frontier was the essence of Americanisation, given that the movement to the west and range of transformations that it operated in a series of settlement waves entailed a progressive departure from European influences. The superiority of white, European settlers seemed to be self-evident due to higher levels of production and economic intensification. Evidently, Turner and his followers failed to see the huge simplification and ideological biases of their teleological claims of democracy, many opportunities and the advantages of privatising the landscape (which had to be first taken from the previous inhabitants). The frontier was considered a field of (white) redemption and great economic success, ignoring the politics of frontier making, the range of sociocultural processes involved, the arrogance of the settlers and the monumental impacts on first peoples, Mexicans, mestizos and other social groups.

The buoyant nationalism proposed by Turner and others is certainly biased in favour of justifying white, capitalist supremacy, but it is right to affirm that the USA, just like Australia, New Zealand, South Africa, etc., continue to be settler societies in which the material and symbolic role of frontiers remain crucial for the organisation of economy and politics. Crosby (2015) defines those areas where land was taken and that became rich food exporters as 'neo-Europes', characterised by new property rights and market-focused production (evidently established at the expense of ecosystems and the condition of indigenous peoples). Weaver (2003) considers that English colonisation, for instance, was obsessed with landed properties that were implemented in settlement colonies, which was an important element of the 'competent' colonial model of Britain (i.e. the English estates worked as models for colonial landowning). According to the last author, the British and Northern European experience of colonialism was superior, more rational and less bureaucratic than the Iberian equivalent, which prevented the expansion of small private properties and led to the accumulation of land in very large estates. However, it seems strange to argue that British frontier-making enterprises were more advanced and wiser than the experiences led by other European powers; in effect, each frontier was unique depending on the local circumstances and politico-economic trajectories, but what fundamentally characterised the unfolding of new frontiers was the coordinated displacement of people and socioecological systems in favour of the exploitation and subjugation of local processes to the demands of core areas. Despite the narratives of freedom and bravery, the extraction of wealth from the North American frontiers, just like in any other equivalent region, benefited from the synergy between defiant private individuals and the state-backed certainties of property rights or even official toleration of audacious acts of land taking.

The second broad group of publications, the diplomatic–legalistic ones, emphasise the (often fuzzy) distinction between frontiers (as a process of economic change, population settling and production integration) and borders or boundaries as lines established by law, treaty, accord or practice that mark the limit of a state or a political unit's territory. In some cases, because of belligerent situations, border regions turn into 'no man's land' (not controlled by any army), demilitarised zones or buffer areas intended to mitigate conflicts. The lines and zones where different states and countries meet have been one of geography's oldest and most enticing issues. Prescott (1987), for instance, provides a number of illustrative examples of old and contemporary border disputes around the world, as in the case of the formation of the modern Greek nation since 1832, when the Greeks began to gain independence from Turkey after three centuries of oppression. National borders have changed or been significantly flexibilised under the expansion of capitalist relations, particularly after the major military campaigns in Europe (the Napoleonic Wars, the First and Second World War), the formation or collapse of economic integration (the expansion of the European Union and the collapse of the Soviet Union) and, more recently, the concentration of power by the China–Russia geopolitical axis, which has repercussions around the globe in terms of geopolitical alignment, opportunities for investment and supply of raw materials (the latter includes South America as an increasingly important source of agri-food commodities and minerals to China). This means that the hyper-dynamic basis of capitalism leads to the creation not only of new frontiers, but also new core areas (as in the case of China's industrial centres). There are numerous Asian examples of the important association between the transformation of forests and agrarian spaces into sites of export-oriented resource extraction and an assemblage of other remote spaces that are transformed into productive sites (Cons and Eilenberg 2019).

Economic frontiers, in turn, are described as areas where any demarcation over which actual or potential mobilities of goods, services, capital, labour, technology or communication flows are low or absent (Pelkmans et al. 2008). The genesis of an economic frontier (which may not coincide with territorialised frontiers) ranges from conflicts, embargoes, natural or political obstacles, protectionism and deliberate policies. The recurrent reopening of frontiers also betrays the stock of contradictions and the fatigue of economic trends in central areas. The frontier helped to deflect domestic political and economic conflicts outward, avoiding the need to confront the protracted problems of inequality and racism. President Trump's border wall—one of the pledges of his reactionary manifesto, presented as a reaction to mounting problems after the 2008 financial crash—represents an allegorical reaction to the historical interconnection of the USA with the rest of the world and a rhetorical attempt to end America's exceptionalism (Grandi 2019). Also relevant are accounts of extractive or resource frontiers, as places where the production and exploitation of resources are organised, including new norms, agencies, contracts and government regulation, to satisfy demands elsewhere (Dressle 2017). A resource frontier is typically dominated by the extraction of natural resources such as timber, minerals, oil and biodiversity goods, which are mobilised after the dispossession of indigenous peoples, squatters or other original inhabitants through complex entanglements of legalities and illegalities that are barely managed by centralised regulatory interventions (Peluso 2018). Resource frontiers show great economic volatility as one resource is exhausted to be replaced by another in a cycle of boom and bust. The organisation of economic frontiers is based on relations of superiority and the imposition of authority that presuppose the privatisation of land and other resources (minerals, timber, water, biodiversity, fisheries, etc.). Capitalist expansion leads to new patterns of resource designation, commodification and exploration, at the same time as it creates new territories defined by the extraction of those resources. A regime of privatised resources is justified on the grounds that it can lead to higher levels of efficiency and improvements through investments of capital and technology, although efficiency claims tend to ignore the destruction of ecosystems and livelihoods and disregard the deep racist foundations of resource frontiers. Because of technological and business developments, new resources are constantly being discovered or 'invented', which results in the continuous mushrooming of resource frontiers across the globe (Kelly and Peluso 2015).

Fourth on our list of disciplinary approaches, the geographical literature, has considered spatial frontiers as settled or sparsely inhabited parts of a country, which are often designated as 'safety valves' for accommodating excess populations of crowded cities or regions (Katzman 1977). Traditional geography, for example Fawcett (1918), maintained the link between territorial formations, boundaries and economic activity as something taken for granted and the result of the 'natural' evolution of productive activities in relation to the availability of resources and infrastructure (e.g. Jones and Darkenwald 1965). More recent works have showed that territoriality is the result of numerous practices, state interventions and sectoral discourses through which the territory exists and achieves institutionalised meanings (Agnew et al. 2008). The manifestation of spatial frontiers is an integral part of the global pattern of uneven development of capitalism that results in the division of labour between areas of production and consumption, as much as the formation of place-based relations and identities (Cox et al. 2008). It is not uncommon to find geographic studies influenced by world-systems theory, especially the theorisation of core–periphery that follows the unity of global capitalism and the single division of labour (cf. Wallerstein 2004). However, one main limitation of such a theorisation is that it totalises capitalism, reducing it to a somehow static interaction between core, peripheral and semi-peripheral states with limited consideration of local cultures and place-based class struggles. For instance, informed by the world-systems analysis, Taylor and Flint (2011: 126) argue that 'the frontier ended with the closing of the world-system [i.e. imperialism, global capitalism], so there are no longer any frontiers—they are now phenomena of history'. This group of authors date the closure of frontiers to around the year 1890, when the territory of the USA was by and large consolidated, including the final incorporation of the last western states, but ignore the important continuities and ramifications in subsequent decades, such as the incessant search for resources, movement of people and the displacement of indigenous groups in different parts of the world (Lamar and Thompson 1981), which basically prove that frontier making remains fully operational.

Such a variety of approaches to frontier and frontier making suggests that in order to address its full complexity, it is advisable to use intellectual tools from more than one discipline in order to account for new rounds of territorialisation, commodification and legitimisation (Ioris 2013). The frontier is a space constantly produced through perennial movement, disputes and reactions, whereas those processes of change are not only social and economic, but intrinsically and profoundly ecological. In other words, the politics of frontier is not restricted to competition between individuals and groups, but it permeates the very understanding of the ecological basis of new spatial settings, as in the case of what can be considered an economic resource and what is allowed to be utilised. All this represents a vigorous call for a politico-ecological perspective that simultaneously and critically interrogates the production of novel spaces, unequal access to nature and the unfair distribution of impacts, attachments to place, questions of identity, interpersonal and interspecific relations and the joint search for alternatives to conventional development (Ioris 2014). This critical politico-ecological standpoint should help to deal with the ecological and resource-oriented basis of contemporary, globalised capitalism and its reverberations across the planet. As described above, frontiers are regions under the influence of, and responding to, the demands of core economic areas, which represent a dynamic process of experimentation and potentially new rounds of resource demand and socioecological exploitation with local specificities. The struggles over land and resources, as well as issues of identity and mobilisation, are largely the product of the specific circumstances of each individual frontier (Peluso and Lund 2013).

A comprehensive and coherent investigative approach is needed to appreciate that the frontier has no rigid borders and no essentialist ontology, but is a space that is being produced after recent incorporation into the sphere of commodity relations and production for the market. The frontiers in a capitalist economy are hubs of intense transformation under the magnetic forces of the centre. The frontspace is gradually transformed into a reflection of what the centre used to be in the past. This means that the difference between core areas (typically the economic capitals of the Global North and main cities of the Global South), frontspaces (strictly speaking frontiers) and frontiers-to-be (areas still beyond the direct influence of capitalist forces, but likely to be incorporated) is more of pattern or gradation than of substance. It is possible therefore to concur with Marx (1973: 459) that 'capital itself creates the conditions for its existence', in the sense that frontier making transforms newly incorporated areas according to the institutions and relations of the core capitalist centres. The frontier is what-is-not-yet but what-is-possible elsewhere. As further theorised by Marx, the 'conditions and presuppositions of the becoming, of the arising, of capital presuppose precisely that it is not yet in being but merely in becoming; they, therefore, disappear as real capital arises, capital which itself, on the basis of its own reality, posits the conditions for its realisation'. The common situation is the one in which economic institutions (such as private property rights, credit and banking, market distribution, transport and communications) are not fully established, and the state has not yet fully installed an apparatus for allocating property regimes, but frontier dynamics end up unleashing the most elemental forces of capitalism, particularly the unrestrained exploitation of labour and nature, the slavery-like control of indigenous populations or migrants, the brutal dispossession of previous residents and unscrupulous rent-seeking behaviour. Because of that, modern landed property cannot be understood and cannot exist 'without capital as its presupposition', and landed property occurs 'merely in its relation to capital' (Marx 1973: 252). Expanding these claims, Luxemburg denounced the capitalist compulsion to encompass the entire globe and treat the world as a store of productive forces to be submitted to market-based laws. 'The existence and development of capitalism requires an environment of non-capitalist forms of production… as market for its surplus value, as a source of supply for its means of production and a reservoir of labour power for its wage system' (Luxemburg 2003: 348).

It is important to emphasise that, despite general trends and shared pressures, geography matters and the frontier is not a passive stock of resources and guaranteed friendly markets. There are no pregiven structures or fixed annihilations of local circumstances—there is no *deus ex machina*—but the frontspace encapsulates a productive dialectic of encounter and continuation. Our main concern here is with the corners of the world, either remote or close, where socio-economic relations and politico-cultural institutions are undergoing a state of flux, rapidly changing and with a high degree of experimentation. There are common trends across the different frontier regions around the planet, as in the case of technology, crop production and social organisation, but new methodological approaches are required to account for commonalities and differences and to capture unique environmental and social patterns (Gradus and Lithwick 1996). The socio-spatial frontier, more than merely a boundary line, is the zone of contact between different socio-economic formations, in a way that the old is being supplanted by new configurations. Evidently, socio-spatial frontiers have never been a wilderness, but are constituted by landscapes previously occupied by indigenous peoples or early migrants, who are normally ignored and displaced due to the advance of the frontier of resource extraction and agricultural production (Ioris et al. 2019). After, and even during, the making of the frontier, the interrelated process of territorialisation takes place, as the re-ordering of the new spatial setting (Rasmussen and Lund 2018). This process of renewal is dynamic and open-ended, in the sense that novel phases of frontier making can supplant previous experiences of resource exploitation or economic renovation. When one change affects an entire frontier site, there are likely to be numerous other changes, including many that are not easily measurable (Rindfuss et al. 2007).

A paradigmatic example is the ongoing encroachment of intensive processes of resource extraction and agricultural production upon the Amazon, which has not only brought millions of migrants to the region but also dramatically inserted economy, ecology and society into the vortex of frontier making (Torres and Branford 2018). This has been a long historical–geographical process of conquest, subjugation, exploration and also symbolic representations of the exoticism of frontiers since the sixteenth century. At the same time that the opening of new regional frontiers intensified in the nineteenth century, romanticism offered representations of pristine nature that were both savage and sublime (Descola 2005). After the colonial exploitation of resources and the collapse of the rubber industry in the early decades of the twentieth century, a new and much more aggressive frontier took shape. This was primarily the result of nationalistic and developmentalist platforms championed by the dictatorial Brazilian state of the post-1964 coup d'état, which comprised intense resource extraction, the installation of a vast physical infrastructure and the expansion of export-oriented agribusiness. The ramifications of those frontier-making plans will be analysed in subsequent chapters, but it is important to emphasise that for more than 40 years the Amazon has been described as a basket of manifold riches and a window to some of the most challenging problems of modernity. The number of articles, documentaries, books and opinions about the conservation and development of the Amazon is bewildering. For example, on 22 November 1981, The New York Times published an article with the title 'The Amazon: A New Search for El Dorado' which declared that 'this last frontier [in Brazil] is beginning to yield to the pressures of politics, commerce and science. The question is no longer whether development is possible, only what form it is likely to take'. In that context, multilateral agencies such as the World Bank influenced national governments to develop a particular understanding of the main economic demands and the vulnerability of Amazon frontiers, including the contrast between settlers and local politicians on the same land, and national policies and economic interests, particularly regarding the allocation of land and construction of roads (e.g. Schneider 1995).

Fig. 1.1 Agricultural areas near Sinop in Mato Grosso (all pictures by the author)

References to frontier in the Brazilian media—which are available in the *hemeroteca* [newspaper library] of the National Library (based in Rio de Janeiro)—illustrate the evolution of the connotative meaning of the word. Between the 1930s and 1940s, there were increasing references to a need to adopt an agrarian reform, to implement regional development plans focused on the western region and to deal with the long-lasting droughts in the northeastern semi-arid region. Those articles were influenced by the policies of industrialisation and modernisation following the 1930 political coup [*Revolução de 30*] that reduced the power of the coffee aristocracy in favour of national integration efforts. The references to frontier changed after the military takeover in 1964, with more specific allusions to the colonisation of the Centre-West and Amazon regions, exploration for mineral resources and the preservation of national security. Yet the media has shown that the Amazon frontier was never open to all in the same way, and not everybody succeeded. With the joint forces of democratisation and liberalisation from the end of the 1980s, there was a shift to a more entrepreneurial language that justified the advance of agribusiness exports and the connection with globalised agri-food and mineral markets (Figs. 1.1 and 1.2 illustrate the scale and technology of soybean production in the region). There have been growing references to the ecological problem and foreign criticism of the lingering devastation of the Amazon biome. Narratives about the Brazilian Amazon have evolved from the rhetoric of national development and integration in the 1970s, to the denunciation of deforestation, resource exhaustion and social

Fig. 1.2 Intense pesticide use required to cultivate Soybean in Mato Grosso

degradation, to the more recent claims that most impacts are, ultimately, justified or have been contained by the implementation of new forestry and environmental regulation. The widespread argument in favour of agricultural productivity and business entrepreneurialism has served a particular politico-economic agenda, but fails to answer the demands of wider society and the requirements to preserve vulnerable ecosystems (Ioris 2017).

Even though it is possible to identify significant shifts and new social experiences, this supposed transition to a 'green frontier' is, at best, partial and vulnerable, without confirmed evidence of real changes in government action and the mentality of decision makers (Schmink et al. 2019). On the contrary, the contemporary economic and regulatory framework in place in Brazil legitimises, stabilises and reproduces agribusiness-friendly policies that depend on the narrative and material relations of frontier making (Ioris 2018). It is evident that the political ecology of the economic frontiers in the Amazon increasingly represent a challenge to society, government and scientists, considering that in the last 15 years, the trend of deforestation has declined (with significant recovery in some years) while agricultural production has increased, but instead of a transition to better environmental governance what actually emerged was a 'governance frontier' in which political and economic power continues to produce new spaces and maintain social inequalities (Thaler et al. 2019). That contradicts the argument of scientists operating within the paradigm of ecological modernisation—the allegation that the economy benefits from moves towards environmentalism, as much as environmental conservation benefits from a more efficient economy—that the frontier is an area 'out there' to be governed, beyond history and existing politics, available to be explored and governed according to various possible approaches, some more sustainable than others. According to the techno-bureaucratic reductionism of ecological modernisation (Ioris 2008), the solution depends on the transition from extractivism or cattle ranching towards a more efficient, conscientious agribusiness organically integrated with conservation goals.

It is becoming increasingly evident that the contradictions and immediatist results of previous periods of frontier making in the Amazon are still vividly present in the contemporary processes of development and regulatory enforcement. For example, soybean exports are not taxed in Brazil, which has become a perverse incentive for further land clearing, while the governments of the states with more soybean production face a very difficult fiscal situation (particularly in the states of Mato Grosso, Goiás and Rio Grande do Sul). Likewise, the 2012 Forest Code introduced a land registry mechanism—the rural environmental registry (CAR)—to monitor and control deforestation, but the practical results have been negligible because of the lack of regulatory enforcement. Research conducted by Galaz et al. (2018) showed that almost two-thirds of the transactions to large beef and soybean-related corporations were transferred through one or several tax havens (including British overseas territories such as the Cayman Islands and the Virgin Islands). Such obscure transactions have not only played a key role in the complex governance of those corporations, but also underpinned environmental changes in the Amazon region. The gap between the legal framework and degradation and tensions on the ground widened significantly after the 2018 election of President Bolsonaro, who came to power with a political manifesto that was formally in favour of reducing environmental and social safeguards. The rise of strong neo-conservative platforms (in Brazil, Italy, USA, etc.) seems to invigorate the prospect of opening new frontiers and expose the old rationalities that always underpinned capitalist frontier making. Without a proper consideration of the ontological basis of those processes, it is difficult to appreciate the interconnected transactions between the expansion of agribusiness in the Amazon and the hidden financial manoeuvres of global capitalism.

In our investigations, it was necessary to combine rich empirical information with the commitment to break new ground in conceptual and interpretative terms (moving beyond the traditional Imperial British boundary-making and other European colonial projects that influenced early works on frontier making and still have reverberations today). Primary data used for the analyses and discussions in the various chapters were obtained during fieldwork campaigns conducted between 2011 and 2018, which comprised interviews, observation and participation in events in towns and settlements across the agricultural frontiers in the Amazon, visits to cropping areas, plantation farms, indigenous communities and peasant communities, contact with private companies, research centres and rural communities. The various campaigns, carried out in different years and locations, focused on specific research questions and became separate chapters in this book (such as agribusiness expansion, family farming, poverty in rich ecosystems, indigenous groups, hydroelectricity and dams.). Interviews and research activities were carried out in Brasília, São Paulo, Rio de Janeiro and Porto Alegre (engaging with representatives of the main agribusiness entities, social movements and corporations) and in the state of Mato Grosso (in the municipalities of Cuiabá [the state capital], Rondonópolis, Sinop, Cláudia, Campo Novo do Parecis, Porto dos Gaúchos, Juína, Lucas do Rio Verde and Sorriso), in the state of Pará (Santarém and Belém) and in the state of Mato Grosso do Sul (Dourados and surrounding municipalities). Local academics, interviewees and informants facilitated the identification of the initial contacts and the research then followed a snowball approach. With the mapping of sectors and organisations, their discourse and stated aims, it was possible to compare intra- and intergroup differences and the range of alliances or disputes (ranging from those strongly against to others fiercely in favour of the prevailing agri-food system among agribusiness farmers, subsistence farmers, urban populations, agro-industrial entrepreneurs, policymakers and politicians, representative agents and the general population). Interviews and meetings were complemented with examination of documents, statistics, websites, leaflets, presentations and newspaper articles found in university libraries and in the archives of public agencies and private organisations. Qualitative materials were transcribed, coded and assessed in Portuguese and, occasionally, other languages; only the extracts reproduced in the book were translated into English. Empirical data were analysed, searching for evidence of the configuration and advance of frontier making, rhetorical and material manifestations of power relations, signs of problems and tensions and subtle evidence of change. All pictures included in this book were taken by the author.

In each of the following chapters, there is a reference to the methodological approaches and intellectual references mobilised to understand different aspects of frontier making. The work involved specific projects, which are deeply interconnected and inform the broader interrogation of achievements and failures. Those projects received support from various sources, which are all warmly acknowledged here, including Coordination for the Improvement of Higher Education Personnel (CAPES, Brazil), the Ecosystem Services for Poverty Alleviation (ESPA) programme, Economic and Social Research Council (ESRC), Arts and Humanities Research Council (AHRC), Newton Fund through the British Academy, Global Challenges Research Fund (GCRF), Mato Grosso Research Foundation (FAPEMAT), Amazonas Research Foundation (FAPEAM), São Paulo Research Foundation (FAPESP) and the Erasmus Programme. The work benefited from workshops, intensive courses and talks in various universities, especially King's College London, Aberystwyth University, Denver University (supported through the Marsico Fellowship), Florida International University, École des Hautes Études en Sciences Sociales—EHESS, Federal University of Mato Grosso (UFMT), State University of Mato Grosso (Unemat), State University of São Paulo (UNESP), State University of Campinas (Unicamp), Federal University of Great Dourados (UFGD), Federal University of Pará (UFPA), Federal University of Rio Grande do Sul (UFRGS), Universidad de los Andes (Colombia) and Instituto Boliviano de Investigación Forestal. A number of governmental and private organisations, including social movements and campaign groups, were contacted and kindly agreed to contribute. One of the most tangible results of that North-South effort was the launch of the Agrocultures Research Network in 2018, funded by the AHRC and coordinated by the author, which includes international experts, national scholars and regional academic and non-academic stakeholders (such as farmer associations, local authorities, community and church organisations, cooperatives, NGOs and artists) to take part in networking activities, discuss existing knowledge and identify gaps in the state-of-the-art of agricultural frontiers in the Amazon.

In order to fulfil the objective of theorising frontier making and also examining recent developments in the Amazon Region, the book was organised into a sequence of interconnected chapters that explore distinct aspects, sectors and areas. After these introductory pages, Chap. 2 presents a reinterpretation of the conceptualisation of frontiers and will argue that capitalism is fundamentally based on 'accumulation by frontier making' where processes of enclosure, production and extraction are recreated and further integrated. The dynamics of frontier making under capitalist relations of production and reproduction will be schematically described as the 'law of scarcity–abundance', as an acknowledgement that human-made scarcities in areas of relatively advanced capitalism require, and prompt, the formation of new economic frontiers where there is a perceived abundance of valued assets. The 'law' is a meta-theoretical synthesis of the gathering of problems in core, 'established' areas in association with promises of solutions in frontier places. It will be claimed that the sphere of capitalist relations serves to expand economic and social activity into newly produced spaces, but retaining an organic and functional association with the centre. One of the most famous theorists and advocates of the positive features of frontspaces was certainly Turner, whose analysis of the expanding American West is well known and widely criticised nowadays. In general terms, his argument is that frontier making was the main promoter of American democracy because of the firm incentives to individualism where land is 'free'; according to Turner, civilisation was encouraged in the large settlement frontier in North America because of the growth of trade. Turner maintained that the USA is primarily defined by the existence and cherishing of western frontiers, that is, the frontier was the most effective instrument of 'Americanisation' as it turned the USA less European and more American; however, it is a specific type of 'new America' that fundamentally mirrors the European epitome. The chapter then revisits what has happened in the state of Mato Grosso, in the southern tracts of the Amazon, which has been at the forefront of frontier making for many centuries, accelerated in the last 30 years with the spiralling growth of agribusiness. Mato Grosso paradoxically achieved the status of a macroeconomic centre, but in practice remains a frontier where abundances and scarcities rapidly follow each other. The state is now considered a highly successful agribusiness experience, but this is a totalizing narrative that disregards mounting contradictions.

Chapter 3 expands the analysis of Mato Grosso introduced in the previous pages through a critical examination of the intricacies of place making in frontier areas. Vast areas of Amazon rainforest and Savannah vegetation have been converted there, since the 1970s, into places of intensive soybean farming basically to fulfil exogenous demands for land and agriculture production. The present assessment goes beyond the configuration of new places at the agricultural frontier, and starts with a qualitative intellectual jump: from place making on the frontier to place making as an ontological frontier in itself. This means that, instead of merely studying the frontier as a constellation of interconnected places, we examine the politicised genesis of emerging places and their trajectory under fierce socioecological disputes. The consideration of almost five decades of intense historic-geographical change reveals an intriguing dialectics of displacement (of previous socioecological systems, particularly affecting squatters and indigenous groups, in order to create opportunities for migrants and companies), replacement (of the majority of disadvantaged farmers and poor migrants, leading to land concentration, widespread financialisation and the decisive influence of transnational corporations) and misplacement (which is the synthesis of displacement and replacement, demonstrated by mounting risks and a pervasive sentiment of maelstrom). Overall, there was nothing inevitable in the process of rural and regional development, but the problems, conflicts and injustices that characterise its turbulent geographical trajectory were all more or less visible from the outset. In the final part, the chapter presents three future scenarios that oscillate between the continuation of current agro-neoliberal trends (facilitated by the appropriation of sustainability and innovation agendas), the return of strong state interventions, or the decline and eventual collapse of the agribusiness sector (due to sanitary risks or politico-economic reactions).

Chapter 4 moves from export-based agriculture to peasant family farming in the socio-spatial frontiers of the Amazon. The ontological features of peasant agriculture are discussed, in particular, the multiple associations with agribusiness. This is an important aspect of the reconfiguration of the Amazon as a vibrant agricultural frontier that has attracted a vast contingent of migrants due to coordinated government plans and, in more recent years, the cultivation of export-oriented crops. One very intriguing feature of this dynamic geography is that small family farming represents the 'other' of capitalist agriculture, but it functions as a hesitant form of alterity that both resists and fulfils agribusiness. In contrast to common arguments, the chapter examines the high entry cost for peasant families willing to move to the frontier: the individual and their family not only have to endure a long journey and difficult circumstances (in the 1970s, it could take weeks to get from the south of Brazil to the Amazon), but also have to cope with the range of values, moralities and identities being shaped; these emerging categories mirror the trajectory in other centres, but assume new configurations (to a greater or lesser extent) at the frontier. Some families had to move several times, from one town or colonisation project to another, until they reached the 'point of commitment', when they ran out of options or there was a decision not to move further. There is a tacit or forced acceptance of circumstances when those who migrated become satisfied with the processes and the local metabolism of scarcities and abundances.

In Chap. 5, another component of frontier making is scrutinised, namely the construction of large-scale water projects, such as hydropower dams and navigation facilities. The Amazon Region is actually one of the most active frontiers of infrastructure expansion, resource extraction and socioecological exploitation in the world today. The Brazilian section of the region has been dramatically inserted into the wider agenda of national development and global trade in recent decades and, since then, negative socioecological impacts have continued to increase, despite the introduction of comprehensive water and environmental legislation at federal and state levels. One of the main processes of politico-economic change is the construction of large-scale water projects, especially for hydroelectricity generation and river navigation (associated with export-oriented agribusiness). In this broad context, the chapter comments on policies and strategies that have encouraged resource grabbing and environmental injustices through an expanded political economy framework that integrates the politics of redistribution and recognition with the politics of resignification. Investments in water projects in the main eastern and southern tributaries of the Amazon River are discussed, particularly to assess the contradictory advance of modernity, the ramifications of corrupted practices and the narrow basis of a prevailing technocratic and reductionist rationality. Following such a trend of modernisation and socionatural reductionism, several river basins are again being incorporated in the agenda of dam construction as critical elements of the next stage of regional development and intensive water management.

Chapter 6 considers how socionatural interactions in the Amazon bear the imprint of old and new forms of injustice and inequality. Amazonian ecosystems end up filled with inequalities and asymmetries that spread from the local to the regional and international scales. In this context, poverty and affluence are metabolised through the appropriation and transformation of forest ecosystems. Poverty is a condition of unsatisfied material and sociopolitical needs caused by combined mechanisms of exploitation, alienation and exclusion. It is a direct product of the prevailing mechanisms of development and environmental management that systematically reinforce hardship and destitution. Governmental instruments and infrastructure investments have attracted different contingents of people to the Amazon, the majority of whom have only marginally benefited from the process of development. The article focuses on geographical areas in the state of Pará, Brazil, and the empirical results demonstrate that the politics of development and the reproduction of poverty in the Amazon do not happen about or around the forest, but with and through the forest. The main conclusions of the research are that development is based on anti-commons strategies, and that poverty is the outcome of political hegemonies exercised over socionature and the systematic failure of poverty alleviation schemes, but also that close interaction with forest ecosystems is central in the production of knowledge and political resistance.

Chapter 7 deals with the impacts of the advance of development frontiers over the land of the Kaiowa, one of the Guarani indigenous groups of South America. Land grabbing and nature commodification have accelerated in the last three decades with the expansion of export-based agribusiness, which has led to the containment of the indigenous population in overcrowded reserves, road encampments and uncertain settlements. As a result, the local rate of suicide is distressingly high, and almost half of the assassinations of indigenous leaders in Brazil have targeted the Kaiowa in the last few decades. However, to the surprise of many, the Kaiowa have shown a latent geographical agency and, despite all the tragedy, suffering, humiliation and terrible neglect by the state, indigenous families have managed to secure small, but precious, territorial victories. Their reaction to frontier making is shaped by religious practices, strong family ties and the ability to internally negotiate the return to their original areas. The wisdom and resistance of Kaiowa-Guarani groups derive from the simultaneous ethnicisation of space and spatialisation of culture underpinning their claims. Overall, frontier making is a muddled situation fraught with puzzles and ambiguities: it is the Kaiowa who are offering real innovation, while agribusiness encompasses an inbuilt obsolescence.

Chapter 8 is about the challenges related to the conservation of the Pantanal area, analysed through a study of the Cuiabá River Basin, one of the areas in the Upper Paraguay River Basin most impacted by the pressures of urban and regional development. The research findings show a clear disconnection between official monitoring data and wider perceptions of processes of ecohydrological change. In addition, it is found that putting environmental regulatory tools into practice is problematic because of the pregiven foundations of conservation strategies promoted by public agencies. The most significant result of the research is the uncertainty in relation to the responsibility for environmental problems, which has created a perverse chain of 'otherness–noneness–nothingness'. This serves to conceal the underlying causes of environmental degradation in the Cuiabá basin and limit possibilities for resolution. The conclusion is that, beyond narrow development and conservation debates, it is necessary to account for a range of highly politicised issues at the intersection between interpersonal relations and socio-economic pressures.

Finally, Chap. 9 consolidates the findings and the argument of the previous sections, returning to the ontological basis of frontier making, insisting that it has always been fundamental for the circulation and accumulation of capital. The perennity of frontier making is not only due to the demand for minerals, land or other resources, or because frontiers represent fresh market opportunities, but crucially because it operates as compensation for the saturation of the existing capitalist relations in core areas. At the frontier, the conventional sequence of time and space is suspended and reconfigured, allowing room for the decompression of tensions and contradictions. Consequently, spatial frontiers function as a mirror, where the most bare and explicit features of capitalism are vividly exposed. This concluding chapter explores the meaning and immanence of spatial frontiers, considering them as a laboratory of historical and geographical agency. It entails a reflection upon the necessity, the configuration and the contestation of spatial frontiers, paying particular attention to the economic and territorial incorporation of the Amazon region and the prospects for political resistance. This final chapter reiterates the main epistemological commitment that pervades the whole book, that is, an intention not to develop an argument against the mere existence of frontiers, but in favour of a centrifugal intellectual exercise that expands the sphere of knowledge beyond the immanence of only one privileged observer, the author (following the advice of Viveiros de Castro 2018). The frontier is a flux of change and contestation, and our interpretative effort needs to humbly acknowledge the intrinsic difficulties of the task. Considering the predicaments of binary Western thinking, Derrida managed to provide helpful and insightful assistance here:

> The discussion becomes interesting once, instead of asking whether or not there is a limit that produces a discontinuity, one attempts to think what a limit becomes once it is abyssal, once the frontier no longer forms a single indivisible line but more than one internally divided line; once, as a result, it can no longer be traced, objectified, or counted as single and indivisible. What are the edges of a limit that grows and multiplies by feeding on an abyss? (Derrida 2008: 30–31)

Before progressing further, it should be acknowledged that earlier versions of some chapter sections originally appeared in the following journal articles: Ioris, A.A.R. 2018. Amazon's Dead Ends: Frontier-making the Centre. *Political Geography*, 65, 98–106 [incorporated into Chap. 2]; Ioris, A.A.R. 2018. Seeding a Narrow Future and Harvesting an Exclusionary Past: The Contradictions and Future Scenarios of Agro-neoliberalism in Brazil. *Futures*, 95, 76–85; Ioris, A.A.R. 2018. Place-making at the Frontier of Brazilian Agribusiness. *GeoJournal*, 83, 61–72 [Chap. 3]; Ioris, A.A.R. 2015. The Production of Poverty and the Poverty of Production in the Amazon: Reflections from those at the Sharp End of Development. *Capitalism Nature Socialism*, 26(4), 176–192 [Chap. 6]; Ioris, A.A.R. 2018. Indigenous Peoples are Collapsing under Agribusiness Frontiers in Brazil. *Cultural Survival*, 20 (December 2018) [Chap. 7] and Ioris, A.A.R. 2013. Rethinking Brazil's Pantanal Wetland: Beyond Narrow Development and Conservation Debates. *Journal of Environment and Development*, 22(3), 239-260 [Chap. 8]. These are reproduced here with permission, and the author greatly appreciates the support received from the various journal editors and referees. I want to thank colleagues at the universities of Cardiff, Edinburgh, Aberdeen, UFMT, UNEMAT, UFRGS, UNESP, UFPA, UFAM, UFRRJ, UFGD, University of Denver, University of Wisconsin–Madison, SOAS, King's College London, University College London and Oxford University. It is also a great pleasure to warmly thank the invaluable help received from a number of farmers, peasants, workers, agronomists, civil servants, urban residents, activists, union leaders, Indigenous and young people who agreed to be contacted, interviewed and spend time with me over the years. Their assistance, patience and cooperation are greatly appreciated. I am, as usual, grateful to my wife Adriane and our son Antônio for their love and support. This book is primarily for the two of you.

References

Agnew J, Mitchell K, Toal G (eds) (2008) A companion to political geography. Blackwell, Oxford

Anderson P (1984) Modernity and revolution. New Left Rev 1(144):96–113

Borges JL (1949) El Aleph. Editorial Losada, Buenos Aires

Cons J, Eilenberg M (eds) (2019) Frontier assemblages: the emergent politics of resource frontiers in Asia. Wiley, Hoboken NJ

Cox KR, Low M, Robinson J (eds) (2008) The Sage handbook of political geography. Los Angeles, Sage

Crosby AW (2015) Ecological imperialism: the biological expansion of Europe, 900–1900, 2nd edn. Cambridge University Press, Cambridge

Derrida J (2008) The animal that therefore I am (trans: Wills D). Fordham University Press, New York

Descola P (2005) *Par-delá Nature et Culture*. Gallimard, Paris
Dressler WH (2017) Contesting moral capital in the economy of expectations of an extractive frontier. Ann Am Assoc Geogr 107(3):647–665
Dobson MJ, Maxfield VA (2005) Roman frontier studies. Liverpool University Press, Liverpool
Fawcett CB (1918) Frontiers: a study in political geography. Claredon, Oxford
Galaz V, Crona B, Dauriach A, Jouffray J-B, Österblom H, Fichtner J (2018) Tax Havens and global environmental degradation. Nat Ecol Evol 2:1352–1357
Gradus Y, Lithwick H (eds) (1996) Frontiers in regional development. Lantham MD, Rowman and Littlefield
Grandin G (2019) The end of the myth: from the frontier to the border wall in the mind of America. Metropolitan Books, New York
Guelke L (1976) Frontier settlement in early Dutch South Africa. Ann Assoc Am Geogr 66:25–42
Guilluy C (2014) Le Crépuscule de la France d'en Haut. Champs, Paris
Ioris AAR (2008) Regional development, nature production and the techno-bureaucratic shortcut: the Douro River catchment in Portugal. Eur Environ 18(6):345–358
Ioris AAR (2013) The value of water values: departing from geography towards an interdisciplinary debate. Geografiska Ann Ser B Hum Geogr 95(4):323–337
Ioris AAR (2014) The political ecology of the state: the basis and the evolution of environmental statehood. Routledge studies in political ecology. Routledge, London
Ioris AAR (2017) Agribusiness and the neoliberal food system in Brazil: frontiers and fissures of agro-neoliberalism. Routledge, London
Ioris AAR (2018) The politics of agribusiness and the business of sustainability. Sustainability 10(5):1648. https://doi.org/10.3390/su10051648
Ioris RR, Ioris AAR (2018) Consolidating the past and risking the future: Colombia's developmental trajectory and the prospects for a lasting peace in the wake of the Havana accord. J Glob South Stud 35(1):155–173
Ioris AAR, Benites T, Goettert JD (2019) Challenges and contribution of indigenous geography: learning with and for the Kaiowa-Guarani of South America. Geoforum, 137–141
Jones CF, Darkenwald GG (1965) Economic geography, 3rd edn. New York, Macmillan
Jordan TG, Kaups ME (1988) The American backwoods frontier: an ethical and ecological interpretation: an ethnic and ecological interpretation. Johns Hopkins University Press, Baltimore and London
Kaltoft P (2001) Organic farming in late modernity: at the frontier of modernity or opposing modernity. Sociol Rural 41(1):146–158
Katzman MT (1977) Cities and frontiers in Brazil. Harvard University Press, Cambridge and London
Kelly AB, Peluso NL (2015) Frontiers of commodification: state lands and their formalization. Soc Nat Resour 28(5):473–495
Kristof LKD (1959) The nature of frontiers and boundaries. Ann Assoc Am Geogr 49(3):269–282
Lamar HR, Thompson LM (eds) (1981) The frontier in history: North America and Southern Africa compared. Yale University Press, New Haven, CT
Lester A (1998) 'Otherness' and the frontiers of empire: the Eastern Cape Colony, 1806-c.1850. J Hist Geogr 24(1):2–19
Luxemburg R (2003[1913]). The accumulation of capital (trans: Schwarzschild A). Routledge, Abingdon and New York
Marx K (1973[1857–1858]) Grundrisse (trans: Nicolaus M). Penguin, London
Miller DH, Steffen JO (eds) (1977) The frontier: comparative studies. Oklahoma University Press, Norman
Pelkmans J, Hanf D, Chang M (eds) (2008) The EU internal market in comparative perspective: economic, political and legal analyses. PIE Peter Lang, Brussels
Peluso NL (2018) Entangled territories in small-scale gold mining frontiers: labor practices, property, and secrets in Indonesian Gold Country. World Dev 101:400–416
Peluso NL, Lund C (eds) (2013) New frontiers of land control. Routledge, London and New York
Prescott JRV (1987) Political frontiers and boundaries. Allen and Unwin, London

Rasmussen MB, Lund C (2018) Reconfiguring frontier spaces: the territorialization of resource control. World Dev 101:388–399
Rindfuss RR, Entwisle B, Walsh SJ, Mena CF, Erlien CM, Gray CL (2007) Frontier land use change: synthesis, challenges, and next steps. Ann Assoc Am Geogr 97(4):739–754
Rosa JG (1956) Grande Sertão: Veredas. José Olympio, Rio de Janeiro
Schmink M, Hoelle J, Gomes CVA, Thaler GM (2019) From contested to 'Green' frontiers in the Amazon? A long-term analysis of São Félix do Xingu, Brazil. J Peasant Stud 46(2):377–399
Schneider RR (1995) Government and the economy on the Amazon frontier. World Bank environment paper number 11. World Bank, Washington, DC
Tanca M (2011) Frontiere, Confini, Limiti: E la Geografia? Between: Rivista dell'Associazione Italiana per lo Studio della Teoria e della Storia Comparata della Letteratura 1(1):1–9
Taylor PJ, Flint C (2011) Political geography: world economy, nation-state and locality, 6th edn. London, Prentice Hall
Thaler GM, Viana C, Toni F (2019) From frontier governance to governance frontier: the political GEOGRAPHY of Brazil's Amazon transition. World Dev 114:59–72
Thorpe B (1996) Colonial Queensland: perspectives on a frontier society. University of Queensland Press, St Lucia
Tolstoy L (2006[1873–1877]) Anna Karenina (trans: Pevear R, Volokhonsky L). Penguin, London
Torres M, Branford S (2018) Amazon besieged: by dams, soya, agribusiness and land-grabbing. Practical Action Publishing, Rugby
Turner FJ (1920) The frontier in American history. Holt, New York
van Lier RAJ (1971) Frontier society (trans. van Yperen MJL). Springer
Verma MK (ed) (2019) Globalisation, environment and social justice: perspective, issues and concerns. Routledge, Abingdon
Viveiros de Castro E (2018) Metafísicas Canibais: Elementos para uma Antropologia Pós-Estrutural. São Paulo, Ubu
Wallerstein I (2004) World-systems analysis: an introduction. Duke University Press, Durham, NC
Weaver JC (2003) The great land rush and the making of the modern world, 1650–1900. McGill-Queen's University Press, Montreal

Chapter 2
Scarcities and Abundances in Place and Time: A Proposed Conceptualisation of Frontier Making

> *But the inchanted island of O'Brasil is not always visible, as those rocks are, nor these rocks have always those apparitions.*
> O'Flaherty (1846: 70)

Abstract This is the main conceptual chapter, which examines frontier making as historically crucial for the circulation and accumulation of capital. The relevance of frontier making is not only due to the demand for minerals, land or other resources, or because frontiers represent fresh market opportunities, but crucially because it operates as compensation for the saturation of the existing capitalist relations in core areas. The dynamics of frontier making can be summarised as the 'law of scarcity–abundance' (LSA), which recognises that human-made scarcities in areas of relatively advanced capitalism require, and prompt, the formation of new economic frontiers where there is perceived abundance of valued assets. LSA is a meta-theoretical synthesis of the gathering of problems in core, 'established' areas in association with promises of solution in frontier places. The State of Mato Grosso, in the southern tracts of the Amazon, has been at the forefront of frontier making for many centuries, accelerated in the last 30 years with the spiralling growth of agribusiness. Mato Grosso paradoxically reached the macroeconomic centre, but in practice remains a frontier where abundances and scarcities rapidly follow each other.

Keywords Scarcity · Abundance · Capitalist relations · Neoliberalism · Financialisation

2.1 Accumulation by Frontier Making

One thing that has certainly been achieved with the aggressive advance of the globalisation of capitalism over the last decades is the creation and accumulation of multiple paradoxes. At first, it seemed that with the success of the Industrial Revolution in the nineteenth century, there was a tangible prospect of overcoming resource and

A. A. R. Ioris, *Frontier Making in the Amazon*, Key Challenges in Geography,
https://doi.org/10.1007/978-3-030-38524-8_2

material constraints; however, the persistence of social, spatial and racial inequalities converged to undermine easy claims of development and civilisation. Significant progress and high standards of living have been reached, but mainly for privileged national elites and subsets of the global human society. At the same time, novel problems relentlessly emerge and grow into large, complicated troubles because of the inability or reluctance to confront the root of the distortions and injustices. Instead of trying to address the deep causes of social–economic tensions, new waves of policy and economic incentives encourage people and practices to be dislocated from one part of the world to another. That has often been the case with unemployment, poverty, land and housing deficits, resource waste and exhaustion, environmental degradation, market saturation and political harassment. The lack of opportunity for some groups and individuals triggers the imagination (of a different reality and a possible better life) but without any guarantee of success (there are none). Uneasy situations, saturated with themselves, are deterritorialised, while dreams and expectations shifted and reterritorialised in new locations. Such attempts to evade and transfer national or location-specific dilemmas to other socio-economic settings seem to be a hallmark of capitalism's unsettling powers and inherent contradictions.

This process of negation and dislocation—what we consider here as key features of 'frontier making'—has been a favourite response to the many of the unresolved contradictions of capitalist society and also a consequence of the search for innovation. For centuries, 'empty spaces' of the American continent, in Australia, Africa, Russia, etc., were the target for the opening of new frontiers in areas considered *terra nullius* through the deliberate and methodical disregard for the existing inhabitants and their socioecological condition. The (partial) mitigation of socio-economic tensions and the quest for novel money-making opportunities in newly opened spatial frontiers have been crucial for the affirmation of capitalist modernity and the functioning of the modern world. Watts (1992: 116–117) considers frontiers as 'particular sorts of spaces' that represent 'the first wave of modernity to break on the shores of an uncharted heartland' with 'their own territorial form of law and (dis)order'. The spatiality of frontier making goes beyond the more immediate relocation from 'core' to 'periphery', entailing a relational interaction and joint processes of realignment and reinforcement that simultaneously happen in old and new areas. In the end, the incorporation of new territories only temporarily alleviates tensions, and without challenging the prevailing relations of production and reproduction. Problems are naturalised, fragmented and depoliticised, new accumulation mechanisms are activated and reinforced, while those who have been most seriously affected by socio-economic developments are compelled to move or risk being blamed for their own difficulties. All that suggests that capitalism is also, and fundamentally, based on *accumulation by frontier making* in which processes of enclosure, production and extraction are recreated and further incorporated in national and global transactions.

One of our central goals is to contribute to the theorisation of frontier making and associate it with the broader politics of scale, that is, the shared experience of capitalist relations of production and reproduction occurring at distinct, but interconnected, scales of socioecological interaction. 'Scale is not necessarily a preordained hierarchical framework for ordering the world—local, regional, national and global.

It is instead a contingent outcome of the tensions that exist between structural forces and the practices of human agents' (Marston 2000: 220). The politics of scale is directly implicated in the production of frontier spaces given that scale is constituted and reconstituted through relations of production, reproduction and consumption interwoven with space. The uneven development of the socio-economic forces of capitalism is essentially multiscale and unfolds through the dynamic political economy of old and new regions (Agnew 2000). Consequently, daily life and class-based differences at the local scale are permeated and help to shape or transgress wider economic and politico-ideological constructs. Likewise, the 'reproduction of the household enterprise is dependent simultaneously on the economic relations of production and on the political relations necessary to protect those relations' (Smith 1989: 24).

In this chapter, we will revisit what has happened in areas where agribusiness advances and reshapes the Amazonia of Brazil, a nation largely shaped by the expansion of internal and external economic frontiers. The next chapter will examine in greater detail the politico-economic activity at the agribusiness frontier, what will be followed by a discussion on infrastructure, peasant family farming, the indigeneity of frontiers and the metabolism of poverty in subsequent chapters. The symbolism of frontier making in the Amazon has pervaded the social and scientific imaginary. More recently, the expression 'agricultural frontier' has also become prevalent and is extensively used by academics and scientists, particularly those working on Brazilian topics.[1] Yet, the ongoing encroachment of export-driven agribusiness into the Amazon region is only the most recent chapter in a long history of the pursuit of new economic frontiers in Brazil. In order to understand the long-term drivers of frontier making, it is essential to consider that the opening of new production areas is not the leftover or the excess of national development, but that it has been central to cultural, political and social change throughout the country. Socio-economic inequalities and socioecological exploitation have also been managed through the proliferation of frontiers and the prospect 'of something better elsewhere' (when 'here' is no longer enough). In that way, the responsibility for problems is shifted back to those exploited and marginalised in the core areas, implying that it would be their own fault if they refused to embark on the journey to a more promising reality at the frontier. Contemporary agricultural frontiers in the Amazon re-enact, once again, the dreams of modernity and prosperity that for generations attracted migrants to the Brazilian west (Ioris 2017a).

To demonstrate the contested politics of scale behind frontier making in the Amazon, our focus will be on agribusiness expansion in the state of Mato Grosso, in the southern tracts of the Brazilian Amazon. Mato Grosso has been at the forefront of economic and political frontier making for many centuries, but the process has accelerated in the last few decades due to the spiralling growth of soybean-based agribusiness. This experience vividly illustrates the paradoxes and extravagances

[1] A search for 'agricultural frontier' on www.scopus.com on 5 December 2018 came up with a total of 499 publications (14% more than in the year before), with more than half (269) dealing with Brazil.

associated with frontier making; Mato Grosso now accounts for around 10% of global soybean production, but the state is an enormous food desert that, like most economic frontiers, still depends for its supply of food on the core, 'consolidated' economic areas in the south and southeast of the country. In addition, the intense economic activity and commodity exports in the Amazon have further moved Brazil towards the periphery of market-based globalisation and reinforced the old pattern of extractivism and socioecological waste (often disguised by calls for efficiency and narratives of sustainable development along the lines of ecological modernisation).

The material presented in the present chapter was accumulated and analysed during research conducted over several years in different rural and urban locations in the north of Mato Grosso, at the Upper Teles Pires river basin, which has become the main soybean production area in Brazil and one of the hotspots of global agribusiness today.[2] Even more revealing, in terms of frontier making, is that the Upper Teles Pires is situated exactly at the transition between forest and Savannah ecosystems. Following Foweraker (1981), interviews, contacts and observations are 'absorbed' into the text and incorporated into the wider analysis without resorting to direct quotations. The study takes on board the recommendation of Pred and Watts (1992: 2) to consider the 'various historical configurations and reconfigurations of capitalism in an effort to understand how difference, connectedness and structure are produced and reproduced within some sort of contradictory global system, within a totality of fragments'. Furthermore, the challenging complexity of the Mato Grosso agribusiness frontier is considered as not only a socio-spatial construction, but also a true analytical tool and a basis for proposing new investigations (Pacheco de Oliveira 2016).

One of the innovative contributions of our investigation is to offer a conceptual framework and an associated reflection on the specific frontier-making experience in order to reconceptualise the wider Amazonian politico-economic trajectory. To achieve that goal, the chapter is divided into two main parts. In the first, after this brief introduction, a theoretical and interpretative perspective is presented, which goes beyond traditional accounts of frontier making in order to emphasise socio-spatial interconnections and interlocked scarcity and abundance (consolidated under the 'law of scarcity–abundance', which will have wider relevance throughout the book, leading to the final discussion and theoretical consolidation in the last chapter). Some authors claim that a new frontier emerges when a new resource is identified or when a new area is cleared, leading to commodification and extraction; however, this is only the visible part of the story: the genesis of the frontier is really in the established politico-economic centre. In the second part, the evolution of the agribusiness frontier in Mato Grosso and, indirectly, in the rest of the Amazon is critically examined, making reference to the roots of the sustained processes of violence, exclusion and hierarchisation that have characterised the long history and contested geography of frontier making in the region.

[2] Teles Pires is one of the two forming tributaries of the Tapajós River, a main affluent of the mighty Amazon.

2.2 The Centrality of Frontier Making and the Law of Scarcity–Abundance

As indicated in the previous chapter, the emergence of new spatial frontiers remains a ubiquitous process in the contemporary world, considering that the decline of frontier making is still an unfulfilled aspect of globalisation and a post-modernist fantasy (i.e. the proclamation of a borderless world, instantly connected and horizontally networked). Novel spatial settings, distinguished by their own patterns of economic production and socioecological organisation, continue to appear, with significant repercussions for national and global societies. One main consequence is that the ontological complexity of frontiers persists as a challenge for social scientists. Imamura (2015) appropriately recommends that the analysis should begin with an inquiry outside academia to observe how the word 'frontier' is used in ordinary speech before it is scrutinised by academics. It is the case that scholars make usage of the word in reference to various types of frontier—political, agricultural, resource, commodity, etc.—and to explain inequalities, separations, gentrification, technological improvements, among numerous other processes. For instance, political frontiers between nations constitute 'markers of identity' and are 'limits of permissible behaviour' perceived in 'different ways by different people' (Anderson 1996: 7). National borders are thus not mere lines on a map, but are a split (even if partial) between contrasting ethnicities, cultures and economies.

In the United States, in particular, the public imaginary is influenced by Frederick J. Turner's persuasive argument about settlement frontiers, basically, the claim that spatial frontiers provided the elemental conditions for freedom and social opportunities in North America (Billington 1963). The understanding of settlement frontiers tends to separate, but also blend together, the socioecological features of 'what was there before' and the material and symbolic practices brought from the original areas where settlers came from. However, as in the case of Turner's, most interpretations seem to miss the multiscale political, social and economic ramifications of frontier making. Scholars have typically described various types of frontier—political, agricultural, resource, commodity, etc.—but fail to take into consideration the range of interests, social differences and political disputes that help to shape frontier spatiality. For example, Demangeon (1932: 636) bluntly considers it 'an exceptional fortune' [*une fortune exceptionnelle*] for a country to have pioneering frontiers and Webb (1952) argues that Western European civilisation was the fortunate result of the opening up of world economic frontiers, which started with Columbus and continued until the twentieth century. Yet, it is rare to find studies that effectively connect local, lived activities with wider politico-economic scales.

Hennessy (1978: 12) rightly observes that frontiers 'have encouraged dichotomies, as they invite Manichean schemes of thought'. For instance, development economists, such as Di Tella (1982), identify frontiers as stocks of untouched resources and sources of wealth waiting to become economically viable (as in the case of 'virgin' land incorporated by the expansion of Germany, Spain, Russia and Portugal). Findlay and Lundahl (2016) likewise explain frontier making in relation to capital investment in land use and in the mobilisation of labour to satisfy the demands of manufacturing in core areas; according to this conceptual model, the frontier is extended and agricultural production increases following initial investments in manufacturing, thus creating a virtuous circle that can lead to the general expansion of the economy. Into the bargain, Alston et al. (1999) emphasise the importance of defining rights to private land to promote market transactions and investments, which will gradually replace the first adventurers with more entrepreneurial farmers. Rationalisations of economic frontiers such as these have been criticised for being naïve, proselytising expressions of environmental determinism and because they ignore the actual actors, historical differences and circumstances involved in frontier making (Mikesell 1960). The main consequence of such economic and institutionalist reductionisms—which broadly follow Locke's philosophy of private property—is the naturalisation of the exploitative and hierarchical basis of frontier making. They ignore the fact that the 'empty spaces' of the American continent, as well as those in Africa, Russia, Oceania, etc., were targeted for the opening of new frontiers—in areas considered *terra nullius*—on the basis of deliberate disregard for the native inhabitants and their socioecological conditions (Richards 2006). As a perverse result, although each frontier location is different, their unstable and mismatched patterns are taken for granted and seem to justify common disdain for their internal configuration.

Contrary to most accounts, frontier making does not happen merely because of the opportunities available in remote new areas, nor does the centre depend on the frontier for its continued existence. The relevance of frontier making derives primarily from the fact that capitalism is not only crisis-ridden but also crisis-dependent—considering that recurrent crises serve to induce changes designed to defend or restore profits (O'Connor 1998)—and new spatial frontiers play a fundamental role in the mitigation and reorganisation of politico-economic structures. Frontiers not only have location advantages and constitute favourable investment areas—what Harvey (2006) termed 'spatial fixes' that minimise the internal contradictions of capitalism—but they are also where old and new socio-spatial features coalesce, resulting in the revitalisation of social and economic patterns of the (relatively consolidated) centre. Regarding the internal tensions of capitalism and its crisis-dependency, O'Connor (1998) suggests that the basic antagonism between the forces and relations of production (originally described by Marx) needs to be complemented with a 'second contradiction' between relations and conditions of production. The second contradiction thesis is one of the main reference points of political ecology and ecosocialism in the critical literature available today. However, it is possible to append a third layer to the twofold contradictions of capitalism described by O'Connor, which is the clash between, on the one hand, the homogenisation of lifestyles, consumption and social attitudes and, on the other, mounting inequalities, politico-economic hierarchies and context-specific problems. In other words, there is a widening gap between the appearance of equality and the crude experience of inequality. This 'third contradiction of capitalism' is profoundly geographical in nature, insofar as exploitation and capital accumulation depend on the imposition of socio-spatial and ideological homogeneity.

It is this third contradiction in particular that frontier making helps to mitigate through the restitution of geographical differences that invigorate capitalist relations (without ever addressing the central irrationality and wastefulness of capitalism). Capitalist development is not only uneven and combined, as Trotsky (1932) emphasises, but capitalist relations are also essentially expansionist and expand in different directions, into different places and practices, to avoid having to confront their own contradictions. The expansionist drive of capital is nothing other than the manifestation of its capacity to creatively reconstruct the world in its own image while avoiding serious challenges to its internal logic (basically, the extraction of surplus value and the commodification of labour and nature). Frontier making is an integral component of capital's intrinsic struggle to suspend socio-spatial dilemmas (including financial cycles, market crashes and socioecological exhaustion—see Solimano 2017), while privatisation, exploitation and accumulation can be expanded to new areas. This stabilising basis of frontier making goes beyond what Luxemburg (1951: 362) considers the insufficiencies of core capitalist areas and the need to spread out to new regions where excessive surplus value can be transformed into productive capital (described as the exploitation of 'territories where the white man cannot work'). In actual fact, frontiers are primarily required because the capitalist centre has become saturated with itself, fraught with uncertainties about its actual legitimacy vis-à-vis its widespread impacts, hesitant about how to cope with past legacies and constantly emerging challenges (migration, urban chaos, food insecurity, social breakdown, to name a few), and consequently needs escape mechanisms to dislocate contradictions to other areas. On the other hand, Luxemburg rightly questioned the idea of rigid distinctions (suggested at the time by people like Kautsky to dismiss the idea that the Soviet revolution could be replicated in Germany) between rich Europe and 'backward' Russia by comparing the complex capitalist regions with the exploitation of Russian agrarian society through a form of capitalism emanating from Western Europe (much in the way that subsequent theorists would describe centre–periphery relations between Europe and its colonies).

To fully grasp the peculiar geographical features of frontier making, it is necessary to subvert any artificial separation between (comparatively consolidated) central areas and (relatively undeveloped) frontiers. It is particularly mistaken to describe frontiers as young areas destined to replicate the politico-economic system of the centre, which, as warned by Massey (2006), constitutes a reduction of geography to history or space to time. Instead of such linear, sequential conceptualisation of the association between centre and frontier, Massey argues that spatial connectedness needs to be considered more seriously, in particular, the understanding that space is the outcome of relations and practices, that it is basically a multiplicity and a constellation of ongoing trajectories. The unique features of centre and frontier must be correctly appreciated, as spatial identities and sociocultural subjectivities are relational and co-constituted through engagement. Based on this, it is possible to infer that frontier activities are distinctive but also predicted in the contradictions and achievements of the centre, just as the centre needs to expand into new frontiers in order to maintain its particularism. Both the frontier and the centre are singular spatial settings but belong to wider processes of space production and its related contestation. Frontier making is a transient suspension of trends and circumstances consolidated in central areas, while also reinstating (in the frontier) similar relations and patterns to those responsible for escalating socioecological problems, economic collapse and inequality in the centre. The new frontiers may be seen as a departure from the centre, but in effect, they are reproducing it. This dialectical mediation between the universal and the specific is resolved by Rancière (2009: 58) in the form of 'the singularised universal' [*l'universel singularisé*], which we can apply to our discussion in order to identify positions of autonomy and individuality located in the interconnected scales from the local to the national and the global (and from the private to the public spheres of interaction).

The main claim so far is that the frontier is not merely an expansion or projection of the centre, but that centre and frontier are permanently connected and interdependent. Frontier making is a systemic process that entails a spatial separation but only through organic interdependencies between so-called 'centre' and 'periphery' (in the Spinozian sense, the frontier is *in* the centre, as much as the centre is *in* the frontier). The frontier ultimately results from the centre being saturated with mounting tensions and multiple contradictions, thus frontier making plays a role in the reduction of socioecological problems and the revitalisation of economic growth and development. We could, therefore, redefine frontier making as a range of interconnected processes of socioecological disruption and spatial reorganisation affecting regions increasingly under the sphere of influence of, and primarily established to attend, national and international capitalist centres. This definition allows us to go beyond a focus merely on the newly incorporated areas and systematically consider interdependencies and continuities. The basic feature of frontier making is the movement of occupation, transformation and landscape discipline in the new zones of appropriation, but tightly connected with the process of internal restructuring in central areas (Moore 2015). Frontier making may have an appearance of chance and chaos, but in practice, it is highly instrumental for the reaffirmation of the 'old' centre through the production of 'new' relations (that are, in effect, projections of the relations established at the centre). Frontier making, therefore, has important parallels with the ambiguous manner in which Thomas More described *Utopia*, a new world that at first sight seems to be the opposite of England but which in practice replicates, in subtle and distorted ways, its many problems.

This first observation leads us to a second crucial postulation: the interconnections between centre and frontier are primarily manifested through the promise of abundance and, paradoxically, the multiplication of scarcity. The interlocking of abundance and scarcity takes place in both central and frontier spaces. Capitalist scarcities, as well as abundances, are human made and produced through historically determined relations, pressures, demands and technologies. Landed property in particular is necessarily relational, violent and political, because land is an economic asset and a source of revenue held against others. At the frontier, the private property of land and resources is established through the imposition, via multiple social and political mediations, of a rational, legitimate institution upon what was considered an anomic, savage and unlawful pre-frontier reality. 'The frontier, which appears as a neutral boundary, serves as a condition of possibility for property's violence' (Bromley 2003: 135), considering that 'violence, law and bureaucracy work in a complementary fashion to mediate the struggle for land on the frontier' (Foweraker 1981:

25). Crucially, because these interconnected spatial settings are gradually inserted into the capitalist logic of accumulation, there are fundamental synergies between selective abundance and widespread scarcity in both the centre and the frontier. No capitalist frontier is ontologically given or predetermined from the outset, but it is possible to anticipate that its unique multifaceted socio-spatial developments will be influenced by abundance and scarcity in the homeland and that it will be marked by the production of new scarcities and abundances at the frontier. These trends are evidently consequences of class-based relations and labour exploitation, which cannot be dissociated from the exploitation of the rest of nature. As examined by Marx, the subordination of nature to capital requires the control and exploitation of labour and this double process of exploitation through capitalist relations connects everyday life with wider, more long-term socio-economic structures.

The classical example is the colonisation of the Americas, which from the sixteenth century involved the disruption of the culture and livelihoods of the Amerindians and the appropriation of land by migrant settlers. Frontier making in the American continent was essentially a 'boom and bust' economy, where, after intense exploitation, resources were depleted (even if only partially) and new frontiers were needed. Particularly the poorest and most marginalised segments of metropolitan society were encouraged, or forced, to migrate to the European colonies under the promise of accessible abundance (land, resources, social opportunities), but the majority of those who migrated would only experience harshness and fresh manifestations of scarcity. Frontier making primarily benefited the bourgeoisie and the political establishment, but to some extent also helped workers in core economic areas, although at the cost of the exploitation of people and resources in the newly incorporated areas (see Bukharin 1929). The experience in the West Indies was paradigmatic, with the decimation of the indigenous population by violence and diseases, and the increasing cultivation of land for sugarcane and cattle (Richards 2006). The removal of 'anti-capitalist tendencies' at the frontier was facilitated by governmental control of land prices and limited access to private property, which compelled immigrants to work long hours for low wages, exploited by the wealthier, capitalist farmers. In the nineteenth century, imperialism expanded frontier-making scarcities in the form of land expropriation and displacement in Asia, Africa and Oceania. After the Second World War, international development was maximised by the Green Revolution and associated with the commodification of even larger tracts of land. Finally, with the advance of agro-neoliberalism in the last quarter of the twentieth century, the intensification of investment in land and agriculture has led to systematic land grabbing.

Therefore, the tendency towards social and economic scarcity in one location is, to a certain extent, compensated—as a counter-tendency of the capitalist system (Mészáros 1995)—by the material and symbolic construction of abundance at the frontier. During more than five centuries of European colonial and imperialist history, the seizure and accumulation of land evolved according to the history and geography of capitalism, and agri-food and natural resource systems were likewise restructured in accordance with the need to renew the accumulation of capital following the incorporation of new economic areas (Dixon 2014). This happened through a powerful politics of scale in which the interaction between local, national and international dynamics was operationalised according to the privileges of a small elite while the majority of the population lived with scarcity. The exhaustion of primitive accumulation in Western Europe led to the colonisation of new frontiers (as in North America), but capitalists strived to contain the relative liberties of economic migrants and had to resort to the 'power of the mother country' and 'use force to clear out of the way the modes of production and appropriation which rest on the personal labour of the independent producer' (Marx 1976: 931). It is important to realise that manifestations of scarcity and abundance take place at different moments in the centre and the frontier. There is a time gap, and when scarcity starts to increase in the centre, there is the prospect of abundance at the frontier. It can be seen that scarcity is never a single process caused by the shortage of means or resources; it is a social relation that unevenly and cyclically alternately affects groups and locations (Ioris 2016a). Certain economic features become scarce, which creates the preconditions for the production of abundance elsewhere. Through the instrumentalisation of scarcity and abundance, many of the worrying tensions that emerge in consolidated areas can be partly released with the movement of (surplus) labour and overaccumulated capital to the frontier.

It is possible to summarise the interplay between (promised) abundance and (widespread) scarcity that underpins frontier making as the *law of scarcity–abundance* (LSA). The LSA is a non-reductionist synthesis of the escalation of problems in politico-economic core areas in association with promises of a solution through frontier making. It is important to note that this is a 'law' in the sociological, non-positivistic sense, that is, the consolidation of social institutions and politico-economic tendencies that result from long-term processes of conflict and cooperation in the history and geography of capitalism. More importantly (I must thank an anonymous referee for their helpful assistance here), the LSA is not an argument in favour of economic determinism or a teleological model of geographical change, but rather a constructivist explanation that takes into account the complex relationship between culture and political economy. The LSA helps to clarify that, while frontier making has the appearance of a centrifugal process (a movement away from the centre), it is in effect mainly a centripetal force, that is, the frontier is instrumental in reinforcing and reinstating the centre. Frontiers happen because of the deferment of scarcity in core areas through the imagined space of abundance projected elsewhere; nonetheless, this promise is normally frustrated and only materialised selectively and at great social and ecological cost. Moore (2015: 87) points out that the 'great secret and the great accomplishment of capitalist civilisation has been to not pay its bills. Frontiers made that possible'. Frontier abundances—as pronounced by Turner (1920), these were at the edge of 'free land' in the American west—derive from the fact that these are considered zones beyond the customary rule of law, therefore lawless and up for grabs. At the frontier, authority is still open to challenge, there are combinations of order and chaos, with the state unable or unwilling to exercise more effective control (Watts 2018).

The various forms of violence that characterise frontier making are rationalised by religious, civilizational or economic discourses that characterise the frontier as a space of inclusion and shared gain (Elliot 2016). 'In the interest of the so-called wealth of the nation, he [the political economist, 'sycophant of capital'] seeks for artificial means to ensure the poverty of the people. Here his apologetic armour crumbles off, piece by piece, like rotten touchwood' (Marx 1976: 932). On the other hand, the perverse and conservative dynamics of the frontier can trigger novel forms of reaction that challenge institutions as soon as they become slightly consolidated. Frontiers are functional for capitalism, but also expose its micro and macro internal fissures, which can lead to smaller or larger forms of resistance. Frontiers are not only places of scarcity, but also sites of potential (Li 2014). An ontological inspection of the frontier reveals a mosaic of disruption and reconstruction, cross-scale interconnections and multiple repercussions of the transposition of values and practices, and collective action based on acute individualism. The frontier is ultimately history through geography and geography with a forward-looking past, as it always brings the old back through the exercise of an uncertain and tentative future. Instead of respecting the role of the subject, the construction of capitalist frontier shatters the possibilities of autonomy both in core and frontier areas. The periphery is the other, the civilizational alter-ego of the centre. That makes any frontier inherently complex and politicised, far from egalitarian. The 'chaos' of the frontier is, in this sense, the offspring of development and of economic problems in the core areas that cannot be easily controlled, let alone harmonised. In the next section, the contested geography of Mato Grosso will illustrate these controversies.

2.3 The Interlocked Frontiers of Mato Grosso

The two main claims presented above—the permanent interdependence between centre and frontier, as well as the law of scarcity–abundance (LSA)—will help us now to examine the general trends of frontier making in the Amazon, from colonial times to the ongoing advance of neoliberalised agribusiness. These two conceptual lenses will be particularly useful in addressing the specificities of the agricultural frontier in Mato Grosso, in the southern section of the Brazilian Amazon, which is shaped by modernisation, commodity exports and global integration demands. The geographical changes in the state of Mato Grosso clearly illustrate the mitigation of socio-economic problems in 'core' national areas and, in addition, the asymmetry of opportunities associated with frontier locations. The 'conquest' of Mato Grosso by agribusiness in the last half-century has been a process of accumulation through frontier making that, in the end, has served primarily to reinforce a highly centralised and exclusionary national society. The appropriation and transformation of Mato Grosso into a perennial economic frontier—still unfolding and even accelerating today—reveals a great deal about the unfairness of Brazilian capitalist modernity and how this is projected across different spatial scales. As explained in more detail below, Mato Grosso is a triple frontier, as it has historically been situated at the margins of Amazonian development, of the Brazilian economy and of the Portuguese empire. These three interlaced frontiers have essentially evolved through the recurrent production of scarcity and (selective) abundance.

The roots of frontier making in the Amazon are to be found not only in Brazilian history, but also in the foundations of Portuguese society and its distinct cosmovision. The drive towards frontier and the persistence of frontier making has long roots in Luso-Brazilian culture. Although Brazil was initially only a small piece of the much broader commercial and colonisation machine, Portugal was itself a country highly marked by frontier making. In effect, since the twelfth century, the Portuguese endured a long and shrewd struggle for independence and political autonomy, which included the need to form strategic alliances with the more powerful European countries and preserve its international maritime trade. Throughout its history, Portuguese identity was widely affirmed in opposition to Spain, which was the perennial 'other' against who Portugal needed to be differentiated (Bastos 1998). Portugal is not only a nation at the margins of Europe, but marked by the contrast between coast and mountains, north and south and, more significantly, its Atlantic position, but Mediterranean socionature (Ribeiro 1963). According to Santos (1993), Portuguese culture is highly heterogeneous and ambivalent, developed at the periphery of European civilisation and prone to navigation. This porous identity and the flexible handling of national interests were highly instrumental in allowing the country to keep its autonomy and independence (particularly regarding the perennial Spanish threat). Following this argument, the high level of malleability made possible for Portugal to maintain its global empire due to the tacit acceptance by the more powerful European countries, which in exchange indirectly explored the Portuguese colonies and partly controlled the economy in the metropolis (vis-à-vis the Methuen Treaty of 1703 that granted tax free to the import of English textiles).

The result is that Portugal was a nation marked by opposition to the neighbouring Spanish kingdom and by the search for new economic frontiers. Santos (1993) further argues that the Portuguese have an 'identity of the frontier' which was sociologically translated into a predisposition towards maritime navigation and global trade. Portugal had already pioneered the route to India around Africa when it decided to claim possession of its South American territory granted by the pope in 1494 under the terms of the Tordesillas Treaty. From the time of the first letter written in 1500 by Vaz de Caminha, the secretary of Cabral's conquering fleet, to the potential. In practice, nonethelessPortuguese King, the new colony was characterised by vast resources and an enormous agricultural, the occupation of Brazil was fraught with violence, ecological degradation, the exploitation of slaves, first peoples and free workers, as well as constant efforts to suppress foreign attacks and local insurgencies. To start with, there was a perennial shortage of workers due to the high cost of African slaves and the recurrent difficulty of enslaving the natives (particularly because of their vulnerability to alien diseases such as malaria and smallpox). More significantly, despite the immensity of the colony, there were persistent complaints about insufficient land and the struggle of poor migrants for survival. Structural land scarcity was obviously caused by the elitist and exclusionary basis of colonial institutions. In the words of Alencastro (2001: 41), 'land and labour [in Brazil] did not come up as independent factors, but as two variables that stem from the driving forces of commercial capitalism'. These were clear manifestations of the LSA, in the sense that impoverished, landless people were led to migrate from Portugal to the colony only to again be deprived of land and to become entangled in the wider processes of colonial exploitation, abject slavery and a hierarchical society.

The growing scarcity of land along the Atlantic coast, together with the prospect of finding precious stones and other riches, was a compelling incentive to explore the vast interior of South America. However, enlarging the colonial Brazilian economy presupposed the transgression of the political border established by Tordesillas. For the Portuguese, most of the land where treasures could be found was beyond reach and could only be accessed through regular incursions into what was formally Spanish territory. 'Indeed, one of the most striking features of South American history has been the success of Brazilian expansion at the expense of its Spanish-speaking neighbours' (Hennessy 1978: 107). One of the most remarkable consequences of such coordinated, though 'illegal', expeditions into the continent was the appropriation and conversion of the majority of the Amazon into a single Portuguese province. The Amazon, a separate administrative unit since the seventeenth century, never existed before the Portuguese invasion and its systematic clashes with the other four competing European powers (not only Spain, but also England, France and Holland). Ultimately, Portugal maintained logistical and military supremacy over the Amazon River Basin, given that the only practicable and easy means of access to Amazonia lay on the Portuguese side of the continent (Prado Jr. 1967). Ever since, frontier making has been a perennial phenomenon in the Amazon, directly associated with cyclical commodity booms. 'In Amazonia, frontiers have not only been opened and closed but reopened and reclosed again and again' (Little 2001: 3).

With the consolidation of the colonial enterprise, it was supposedly possible to transplant the national 'human wealth' and recreate Portuguese civilisation abroad (Serrão 1980: 98), although in practice it meant the reproduction of the rigid political hierarchy and exclusionary social relations that defined Portugal's society. Freyre (1961: 23) famously argued that Brazil has 'a predominantly Portuguese culture in its characteristics' manifested in 'processes, methods and techniques of blending and harmonising European civilisations with the tropical landscape and human cultures of this part of America'. The Portuguese colonisation effort around the world has been often described as the *Pax Lusitana* and that can give rise, according to Freyre, to a specific field of study, the 'Luso-tropicology'. However, instead of a 'peaceful' *Pax*, the occupation of Brazil was marked by cruelty and repression, considered necessary to sustain labour exploitation and guarantee territorial integrity. It all depended on the violent assault and cultural obliteration of the native population and the miserable impoverishment of the survivors (Hemming 1987). From the perspective of the indigenous population, the advance of the frontier represented absolute scarcity, that is, the denial of the very possibility of social existence and, for the majority, of physical survival. The process has predictably followed the overall logic of perceived abundance that recreates material scarcity due to pillage, wastage and socioecological disruption (Heckenberger 2005).

In that context, the history of what is now Mato Grosso started in the beginning of the eighteenth century, with the increasing presence of Brazilians and Portuguese involved in the extraction of diamonds and gold in what was officially until then Spanish territory. The foundation of the first settlements by Portuguese–Brazilian explorers gradually displaced the activity of Spanish missionaries and weakened the influence of Asunción and Santa Cruz de la Sierra (Lucidio 2013). New political borders were agreed later in the century by a series of international treaties that granted Portugal the possession of much larger tracts of land (most of it was previously assigned to Spain by the Pope), approximately the size of contemporary Brazilian territory. After national independence in 1822, the Brazilian government promoted fierce policies of national unity and containment of the numerous rebellions. The Regency period (1831–1840) was particularly turbulent and there was a real prospect of territorial fragmentation because of accumulated socio-economic tensions and a strong sense of political contempt among regional elites. It was a long and painful road to national unification, in which scarcity and abundance were managed by able political leaders on behalf, primarily, of land and slave owners. One key feature of nation-building, according to Santos (1993), was that Brazil internalised centre and periphery in order to keep the territory united. It was an astute plan that involved the consolidation of international political borders and the subordinate integration of national economic frontiers. Although 'Mato Grosso never became a center of more than secondary importance' (Prado Jr. 1967: 78), its relative significance increased with the outbreak of a large-scale war (1864–1870) that followed the invasion of Brazilian territory by Paraguayan troops. After the war, Mato Grosso was again reduced to a mineral and agricultural periphery, with some incipient farming activity and only marginally involved in the 'rubber boom' that dominated the Amazon economy at the turn of the twentieth century (Wilcox 2017).

The decline and eventual collapse of the rubber activity, after 1912 (due to British biopiracy and the production of latex in Asia), affected Mato Grosso much less than the core production areas in the centre of the Amazon, but further contributed to its marginal condition in the national economic geography. As observed by Bunker (1984), extractive economies are deeply dependent on the demands from the politico-economic centre and the result is a succession of discontinuities in time and space. In the first decades of the last century, Brazil was a young republic (since the removal of the last king of the House of Braganza in 1889) and the nation was approaching its first centenary of independent history (achieved in 1822), but different than the rest of the Spanish-speaking continent, Brazil remained united after the Portuguese left and even managed to conquer more land from the neighbours (notably the acquisition of Acre from Bolivia in 1903). Nonetheless, there was within its borders a mosaic of poorly integrated local economies and untouched (from the perspective of the national elite) reserves of natural resources. Much more than half of the country was still considered primitive, removed and frozen in the past, but holding wealth and opportunities needed in the politico-economic centre. It was the paradox of territorial integrity and largely peaceful borders containing inside an economic and social frontier. In that connect, the advance of the economic frontiers revealed a great deal about the achievements and shortcomings of both frontier making and, more generally, of the deep national problems.

Despite some infrastructure and operational improvements, the economic situation of Mato Grosso remained a cause of significant embarrassment during the first decades of the twentieth century. While the south and southeast of Brazil were gradually becoming more urban and industrialised, in Mato Grosso, there were only meagre signs of agricultural intensification and economic growth. The process of frontier making had operated more or less unabated for several centuries based on the prospect of abundant minerals, land and resources, which corresponded to short-lived periods of prosperity quickly succeeded by the more enduring condition of isolation and renewed forms of scarcity. Mato Grosso was at the periphery of a peripheral country that was at the time highly dependent on the export of coffee and a few other primary commodities. The main economic frontiers at that point were located in the west of the State of São Paulo, which offered the best socio-economic prospects for the poorer segments of society. With a limited domestic market and scant capital investment, one of the only options available for the majority of the population was internal migration, although this necessarily represented the 'sacrifice of certain zones in favour of others' (Sodré 1964: 313). Furtado (2006) demonstrates the concurrence of scarcity and abundance underpinning the national economy of the time. According to the last author, after the stock market crash in 1929, the price of the main Brazilian commodity (coffee) suffered a serious deterioration, although production continued to increase (peaking in 1933); excess coffee production was caused by a stagnant market and the country was forced to adopt radical anticyclical policies, including the physical destruction of stocks. Additional economic strategies started to focus on import substitution industrialisation, emphasis on the domestic market and greater economic integration of the different parts of the country.

In the 1930s, through resolute interventions by the federal government, the Centre-West region was targeted to become the main development frontier. The 'March towards the West' was a government programme launched in 1938 in a context of national mobilisation and economic expansion promoted by the authoritarian Vargas administration (1937–1945). The March involved exploratory surveys, road construction, agricultural colonisation and various economic incentives. The intention was to integrate and modernise the Centre-West, but the ultimate long-term goal was to prepare the terrain for the future economic incorporation of the Amazon. Mato Grosso, located both in the Centre-West and in the Amazon, then became a Holy Grail of land speculation, fuelled by the easy, almost cost-free, concession of public land to private petitioners (Moreno 2007). In the post-World War II period alone, the government of Mato Grosso indiscriminately titled around 4.2 million hectares of public land (Garfield 2001, in Jepson 2006). In the 1950s and 1960s, while Mato Grosso's vast territory was being partitioned among farmers and speculators, central economic areas in the eastern states were marked by escalating social conflicts and calls for labour reform and agrarian legislation. It was a situation of mounting scarcity felt by the majority of the population which contrasted with the concentration of wealth resulting from industrialisation and other developmentalist policies. Instead of seeking inclusive and democratic solutions to these fundamental challenges, the long Brazilian tradition of political shortcuts prevailed once again. In 1964, a military coup, expected and supported by the United States, instituted a long dictatorship (it lasted 21 years) that worked in favour of hyper-conservative modernisation along the lines of state gigantism and technocratic developmentalism.

To satisfy public opinion, the generals even legislated on agrarian matters—notably, the 1964 Rural Land Statute—although this was a bureaucratic, anti-reform response did not address structural problems. Their main agrarian strategy was actually the reinforcement of frontier making in Mato Grosso and the rest of the Amazon. A vast agricultural frontier was established in the 1970s by the military government, which basically reinstated the same ideological orientation adopted in previous decades under the March towards the West (Velho 2009). Instead of food production or rural development goals, what actually happened was the formation of a 'frontier of landed property' in state-owned areas. It was a process loaded with ideological and political claims, according to the long-established logic of promising abundance to overcome scarcity. Such massive internal migration closely observed the LSA, given that impoverished groups in the south and northeast were compelled to move to colonisation projects in the newly opened frontiers, only to struggle with new rounds of scarcity and, in many cases, were forced to migrate again further inland. The Amazon was particularly targeted for the relocation of 'troublemakers', that is, landless groups who were demanding, within the limits of a brutal military dictatorship, land for workers and peasants. Especially in the southernmost state of Rio Grande do Sul, farming areas established by the migration, more than a hundred years earlier, of peasant families from Italy, Germany and other European countries were suffering from acute land scarcity caused by the unequal agrarian structure, aggravated since the 1950s by the advance of agribusiness in the context of the Green Revolution. Many impoverished families were hurriedly transferred to the Amazon and, as a rule, found it very difficult to settle, produce and survive because of various economic and social adversities, which soon led to the abandonment and concentration of land. Relevant here is the observation by Sato (2000) that frontiers are actually in-between spaces that are occupied by different groups of in-between people.

In parallel with colonisation schemes involving peasant families, private individuals and companies could acquire land from public authorities in order to secure property rights and then claim public subsidies and related benefits. If land along the Transamazon Highway in the state of Pará was the priority frontier for settling peasants in public colonisation projects, Mato Grosso was the main arena for entrepreneurial resettlement.[3] Application for land titles was massively tainted by corruption and happened without even the most basic attention to circumstances on the ground, topographic features, local settlements or even other documents issued for the same area (Ioris 2017b). Land allocation typically involved the mere drawing of lines on a highly imprecise and crude map by government officials working hundreds of kilometres away from the area. The result was that the expansion of private property had very limited agricultural results, but it was nonetheless legitimised by the sheer abundance of land and, more importantly, generous government assistance and rural credit (Schwantes 1989). As in other countries, the new farmers were

[3]The focus on larger properties in Mato Grosso (under the 1974–1979 government of General Geisel) came a few years after the largely unsuccessful resettlement of peasants to small-scale properties in the state of Pará (during the 1969–1974 administration of General Medici).

more like speculators with no interest to work in the land as a personal enterprise or buildup a long-term tenant estate, instead purchasing large sections of unimproved land, intending to sell after land values had risen sufficiently to make their sale remunerative (Bogue and Bogue 1957). In total, the production of Mato Grosso's agricultural frontier involved more than 100 resettlement projects and incorporated around four million hectares between 1970 and 1990 (Jepson 2006). Particularly in the Upper Teles Pires river basin in the centre-north of Mato Grosso, land settlement was encouraged by the construction of a federal road, the BR-163, by the Brazilian army (Bernardes et al. 2016). Later, after intense timber extraction, this region became the main national centre for soybean production.[4]

Mato Grosso's land allocation experience deviated somewhat from the classical frontier-making model described in the literature, which hypothesises the initial arrival of the first pioneers, typically subsistence farmers and adventurers with little to lose, who are then gradually replaced by more commercial farmers and private companies, leading to various degrees of tensions between the different groups. In Mato Grosso, there was no spontaneous transition from 'farming' to an 'agricultural sector' (as dealt with by Bernstein 2010); what actually happened was the writing off of the socioecological circumstances and the imposition, through top-down state action, of exogenous practices and bureaucratic goals informed by values and priorities decided elsewhere. The aggressive mobilisation of land and resources to create a large-scale agricultural region was triggered by the political and ideological vision of the federal state (especially after the 1964 military coup) on behalf of a conservative vision of modernity. The government championed national development and integration plans (financed largely from multilateral loans fuelled by petrodollars) that triggered internal migration of farmers and fostered new rounds of primitive accumulation by private companies entitled to receive generous subsidies once they started operating in the Amazon. This is related to the claim by Hung (2014) that frontiers are based on simultaneous dualities (e.g. inclusion/exclusion, periphery/connection, modernity/primitiveness) that constitute essential dilemmas through which the state can 'tailor' different meanings to meet contingent market demands.

Rather than bringing development to an area considered destitute (as hypothesised by Bauer 1991), frontier making in Mato Grosso was speculative and aimed to mitigate the scarcity of land and limited social opportunities elsewhere in the country. As a result, there were serious doubts about how resilient this agricultural frontier would be without sustained and generous state assistance. Because of the gradual exhaustion of the frontier-making model introduced in the 1970s, which was highly dependent on public funds and direct state intervention, some authors concluded that it had become a decadent frontier and even anticipated the retreat of capitalism

[4]Interestingly, the Upper Teles Pires is a frontier within the wider frontier, as it is situated at the uncertain boundaries between Savannah [*cerrado*] and the Amazon rainforest (the former has lower legal protection compared with forested ecosystems). The fuzzy frontier of the forest has been used by the farming community as an important excuse to clear more land and claim their 'environmental prerogatives' (some even call it their 'green passport'); a typical narrative is exemplified in this interview with a municipal authority: 'we protect the forest because we mainly cultivate land previously occupied by the Savannah'.

from the Amazon (Cleary 1993). This model, championed by the Brazilian military, proved to be too costly, scandalously inefficient and obsolete, which became evident after the end of the dictatorial regime in 1985 and the subsequent fiscal crisis. Nonetheless, after a turbulent period of adjustments, and to the surprise of many, the agricultural frontier re-emerged and flourished from the late 1990s onwards, under a favourable convergence of agrarian reorganisation (i.e. further land concentration), global market opportunities (the commodity boom in the 2000s) and macroeconomic liberalisation policies (centred on monetary stability, global trade, financial speculation and deindustrialisation). This corroborated the claim by Pichón (1997) that land use at the frontier of agricultural development and colonisation is always complicated and tends to be adaptive to new circumstances.

In the last two decades, Mato Grosso has transformed into one of the main agribusiness production hotspots in the planet, particularly fuelled by the strong demand for soybean from Asian markets. While the Brazilian economy has rapidly deindustrialised and the public deficit soared, Mato Grosso's agribusiness is one of the few economic sectors in the country able to demonstrate high levels of efficiency and significant contributions to trade surplus. Soybean exports have been particularly important because they provide foreign currency, which helps to stabilise the national economy and reduce the public deficit. The agricultural frontier, originally opened by Vargas and expanded by the military, has been reinvented as 'corporate Amazonia' today under the sphere of influence of land investors, private banks and transnational corporations (not only the big international names, Cargill, Bungle, ADM and Dreyfus, but also impacted by new Brazilian TNCs, such as Amaggi, BRF, Marfrig and JBS). Mato Grosso is now, for the first time in its history, praised as an example of economic success and entrepreneurial efficiency. If one listens inattentively to the narrative of production efficiency and economic triumph—repeated daily by political leaders and echoed through the mass media—one would probably come to the conclusion that Mato Grosso now occupies the centre of gravity in the Brazilian economy and political landscape. However, the acclaimed victory of Mato Grosso's agribusiness is actually an '[interlocked concept' that preserves the perverse basis of frontier making. Despite its crucial economic role and the growing influence of regional political elites, Mato Grosso never stopped being a frontier. More than that, frontier making never ended in Mato Grosso, but remains a perennial necessity, much more than a simple contingency.

Regardless of discursive and symbolic constructions, the current state of affairs in Mato Grosso organically depends on new and more sophisticated rounds of frontier making. First of all, the defensive rhetoric repeatedly employed by agribusiness leaders and their political and academic allies—for instance, the common claim that 'Mato Grosso is the Brazil that is doing well' [*o Brazil que dá certo*]—is in itself an indication that relations are not yet settled, but that the region is still undergoing a process of socio-spatial consolidation. Second, Mato Grosso's export-oriented agribusiness remains firmly subordinate to financial, technological and political centres that determine what and how commodities should be produced and commercialised. Third, because of growing market demands and mounting socioecological degradation, as well as the rising price of rural land, the agribusiness sector is constantly in search of new production frontiers deeper inside the Amazon region. Fourth, Mato Grosso is an economic frontier in which abundance and scarcity are organically connected and follow each other. Neoliberalised agribusiness, since the late 1990s, has revived the promise of abundance, but even more rapidly restricted socio-economic opportunities. According to the state and national statistics, 44% of the population lives in a situation of social vulnerability and 18% below the poverty line (Circuito Mato Grosso 2016). Agribusiness accounts for around a third of state GDP, but social inequality and the quality of public services have been seriously affected by widespread corruption, which affects almost the whole political system and has led to the imprisonment of many corrupt authority figures, including a former state governor (El País 2017). One of the most tragic results is the growing urban periphery, with abject levels of poverty, which is also a problem in the towns and villages in the main agribusiness areas. Increasing land prices over the last two decades have limited the possibility for smaller farmers to acquire their own property, creating incentives for intrastate migration and for further deforestation and associated environmental degradation (Ioris 2016b).

The specific conditions for agribusiness expansion in Mato Grosso ultimately reproduced the monopoly of land-holding—a dwindling number of increasingly wealthier landowners—that characterised previous rounds of frontier making. This situation of selective abundance and shared scarcity is further reinforced at the local scale of farms and production units. Frontier making involves a dialectic between unique socio-spatial experiences and strong homogenisation pressures. There is growing homogenisation of production, given that almost all farms use the same technological package (GMO soybeans, large machinery, digital technology, agrochemicals, etc.),[5] which has resulted in biodiversity erosion, water pollution and soil degradation. These socioecological impacts are all manifestations of growing scarcity. In addition, technological uniformity corresponds to a politico-ideological homogenisation, as it is not easy to find public spaces to question and criticise regional development trends (even in the local universities). The associations that represent the agribusiness sector (e.g. FAMATO and Aprosoja) are centralised organisations with a façade of public participation but subservient to the politicians and large-scale farmers who virtually control the public sector of Mato Grosso. One notable example of the scarcity of alternative thinking is the appropriation of environmental regulation by the agribusiness sector, along the lines of ecological modernisation (Baletti 2014, describes it as the 'greening' of soybean production via environmental governance). In this way, the agricultural frontier not only encroaches upon ecosystems, but also gradually appropriates the practice and rationale of environmental law, obviously focusing on market-based solutions (e.g. carbon trade, compensatory measures for past degradation in a location different than the one impacted, deforestation amnesties, etc.).

In the end, agribusiness, particularly the neoliberalised version that now prevails in Mato Grosso, has never undermined or reduced the importance of capitalist frontier

[5]Mechanisation, digital equipment and other related technologies help to control labour (a perennial problem in agriculture) and facilitate capital investment in farming.

making. On the contrary, frontier making continues to serve the stronger interests of the Mato Grosso elite (such an elite is itself the product of frontier making) and the hegemonic demands of core political and economic centres elsewhere in the country (Brasília and São Paulo, above all). A supposed post-frontier stage, as announced by some scholars, has not materialised in the region, as frontier making remains predicated upon the need to compensate for national and local tensions. Just like diamonds, rubber and cattle in the past, the soybean-based economy of Mato Grosso persistently faces many challenges (aggravated by serious ecological degradation and global warming), but without fundamental politico-economic transformations, new cycles of frontier making are likely to continue to ensue. In that context, the national state—as the key neoliberal player (Ioris 2012)—maintains a leading role in frontier making, both through the facilitation of processes of land grabbing, utility privatisation and agriculture financialisation, and through the flexibilisation of social and environmental regulation.[6] Neoliberalism, as a class-based attempt to reorganise production and re-enact exploitation, proves to be an institutional and moral frontier of capitalism and, at the same time, encourages the production of new spatial frontiers.

2.4 Conclusions

Frontier making has been an integral component of the evolution and renovation of capitalism, not merely because of the conquest of territories and resources, but because of the strategic interactions between old and new areas. The advance of capitalist relations of production and reproduction continues to rely today on accumulation by frontier making, as illustrated by the persistent formation and constant reinforcement of socio-spatial frontiers in the Global South. The main reason is that the process of frontier making simultaneously connects and transforms both the core and periphery circuits of capital. Marx and Engels (1970: 88) even argued that settlement frontiers, as in the case of North America, served to accelerate the progress of capitalism as the settlers bring ‘advanced forms of intercourse’ that they were not able to establish in the old (European) countries, which is a phenomenon that will eventually impact the original areas from where the settlers departed. There is, therefore, a totality of relations that underpins capitalist frontier making and that connects unique socio-spatial circumstances with the wider complexity of capitalist processes of production, reproduction and legitimisation. Frontier making persistently happens because of the overflow of class-based disputes in geographically ‘consolidated’ areas and the specific opportunities to establish novel socio-economic mechanisms at the frontier. The unique space of each frontier is what Pred and Watts (1992) describe as ‘multiple modernities’ consistent with ‘multiple capitalisms’. In addition, Li (2014)

[6]In Gramscian terms, the interpretation of frontier making needs to be firmly situated in the political ecology of the state (Ioris 2014) in order to understand the multiple ramifications of the power asymmetries that have helped to shape the frontier, including this movement from ‘nature as resource’ to ‘nature as business opportunity’.

perceptively points out that capitalism is not simply a totalising system that incorporates everybody, and that the adoption of capitalist forces in frontier regions is not an inevitable phenomenon, but full of conjunctures (i.e. subject to various conditions).

Capitalist frontier making was historically initiated as part of the European expansion around the world and over time, it became a fundamental pillar of Brazilian political, social and economic organisation. Following the law of scarcity–abundance—which encapsulates the overall tendency to deal with mounting scarcity in central areas through the promise of abundance at the frontier—families and groups have, for many generations, departed from areas fraught with (human made) scarcity, attracted by the promises of abundance and a better life elsewhere in the country, as in the case of Mato Grosso (located in the southern section of the Amazon biome, at the transition between Savannah and forested ecosystems, and renowned for its abundance of land, water and biodiversity). According to the intertwined pledges of abundance and the more regular delivery of multiple scarcities, the expansion of Mato Grosso's economic frontier has deliberately been used to help regulate long-term, unresolved socio-economic contradictions in the national economic centres. In that way, the perverse configuration of Brazilian society, especially and prominently characterised by acute inequalities and abject gulfs between poor and rich groups, was reinforced by the recurrent instrumentalisation of frontier making. Frontier making in Mato Grosso also contributed to the spatialisation of class struggle at different, nested scales of interaction: from local disputes over land and resources, to nation-building and global flows of commodities and geopolitical concerns.

Frontier making was not only important during colonial and early independence periods; the modernisation of Brazil unremittingly depended on the perennial production of new frontiers. It remained a crucial escape mechanism needed to alleviate socio-economic and socioecological tensions. As a consequence, Mato Grosso was converted into a triple frontier—at the periphery of Portugal, Brazil and the Amazon region—and today it still functions as a technological, ethical and political frontier, despite the fact that it currently contributes much more significantly to the national economy via the export of agribusiness commodities. The process of frontier making was never interrupted or weakened, even now when Mato Grosso occupies a prominent position in the national politico-economic arena. Mato Grosso continues to be a zone of experimentation, migration and reconstitution, where identities and allegiances are still shifting and the democratic rule of law has not yet materialised. The frontier was always promoted by the national state but the same state never had much interest in guaranteeing a socially and spatially inclusive institutional order to resolve land-related conflicts. Formal laws are diluted and largely replaced by other codes marked by pragmatism and etched by the balance of power between different social groups. For all these reasons, frontier making in Mato Grosso is both a concrete experience and also a metaphor for the uncertainties created by the geography of capitalism in Brazil and the constant need to provide (albeit circumstantial and transient) responses to the problems accumulated in core and newly produced spatial areas.

Acknowledgements This chapter is based on an article by the author as published by Elsevier: Antonio A.R. Ioris, Amazon's dead ends: Frontier-making the centre, Political Geography, Volume 65, 2018, Pages 98–106, ISSN 0962-6298, https://doi.org/10.1016/j.polgeo.2018.05.011.

Reprinted with the kind permission of Elsevier.

References

Agnew J (2000) From the political economy of regions to regional political economy. Prog Hum Geogr 24(1):101–110

Alencastro LP (2001) O Trato dos Viventes: Formação do Brasil no Atlântico Sul. São Paulo, Companhia das Letras

Alston LJ, Libecap GD, Mueller B (1999) Titles, conflict, and land use: the development of property rights and land reform on the Brazilian Amazon frontier. University of Michigan Press, Ann Arbor

Anderson M (1996) Frontiers: territory and state formation in the modern world. Polity Press, Cambridge

Baletti B (2014) Saving the Amazon? Sustainable Soy and the new extractivism. Environ Plan A 46:5–25

Bastos C (1998) (In)visible borders: ideologies of Sameness and otherness in a Portuguese context. Port Lit Cult Stud 1:19–32

Bauer P (1991) The development frontier: essays in applied economics. Harvester Wheatsheaf, London

Bernardes JA, Buhler EA, Costa MVV (eds) (2016) As Novas Fronteiras do Agronegócio: Transformações Territoriais em Mato Grosso. Lamparina, Rio de Janeiro

Bernstein H (2010) Class dynamics of agrarian change. Fernwood Publishing, Halifax and Winnipeg

Billington RA (1963) America's frontier heritage. University of New Mexico Press, Albuquerque

Bogue AG, Bogue MB (1957) "Profits" and the frontier land speculator. J Econ Hist 17(1):1–24

Bromley N (2003) Law, property, and the geography of violence. Ann Assoc Am Geogr 93(1):121–141

Bukharin N (1929[1917]). Imperialism and world economy. Martin Lawrence, London

Bunker SG (1984) Modes of extraction, unequal exchange, and the progressive underdevelopment of an extreme periphery: the Brazilian Amazon, 1600–1980. Am J Sociol 89(5):1017–1064

Circuito Mato Grosso (2016) 44% da População de Mato Grosso Vive em Vulnerabilidade Social. http://circuitomt.com.br/editorias/cidades/91234-44-da-populaaao-de-mt-vive-na-linha-da-pobreza.html

Cleary D (1993) After the frontier: problems with political economy in the modern Brazilian Amazon. J Lat Am Stud 25(2):331–349

Demangeon A (1932) Pionniers et Fronts de Colonisation. Annales de Géographie 41:631–636

Di Tella G (1982) The Economics of the Frontier. In: Kindleberger CP, di Tella G (eds) Economics in the long view. Palgrave Macmillan, London, pp 210–227

Dixon M (2014) The land grab, finance capital, and food regime restructuring: the case of Egypt. Rev Afr Polit Econ 41(140):232–248

El País (2017) Mato Grosso, o Estado Rico Onde os Escândalos Políticos "Monstruosos" se Sucedem. https://brasil.elpais.com/brasil/2017/10/24/politica/1508811228_204258.html

Elliot H (2016) Planning, property and plots at the gateway to Kenya's 'new frontier'. J East Afr Stud 10(3):511–529

Findlay R, Lundahl M (2016) The economics of the frontier: conquest and settlement. Palgrave Macmillan, London

Foweraker J (1981) The struggle for land: a political economy of the pioneer frontier in Brazil from the 1930s to the present day. Cambridge University Press, Cambridge

Freyre G (1961) The Portugueses and the tropics. Lisbon

Furtado C (2006) Formação Econômica do Brasil. São Paulo, Companhia das Letras
Harvey D (2006[1982]) The limits to capital. Verso, London and New York
Heckenberger M (2005) The ecology of power: culture, place, and personhood in the Southern Amazon, A.D. 1000–2000. Routledge, Abingdon
Hemming J (1987) Amazon frontier: the defeat of the Brazilian Indians. Papermac, London
Hennessy A (1978) The frontier in Latin American history. Edward Arnold, London
Hung P-Y (2014) Frontiers as dilemma: the incompatible desires for tea production in Southwest China. Area 46(4):369–376
Imamura M (2015) Rethinking frontier and frontier studies. Polit Geogr 45:96–97
Ioris AAR (2012) The neoliberalization of water in Lima, Peru. Polit Geogr 31(5):266–278
Ioris AAR (2014) The political ecology of the state: the basis and the evolution of environmental statehood. Routledge, London
Ioris AAR (2016a) Water scarcity and the exclusionary city: the struggle for water justice in Lima, Peru. Water Int 41(1):125–139
Ioris AAR (2016b) The paradox of poverty in rich ecosystems: impoverishment and development in the Amazon of Brazil and Bolivia. Geogr J 182(2):178–189
Ioris AAR (2017a) Places of agribusiness: displacement, replacement, and misplacement in Mato Grosso, Brazil. Geogr Rev 107(3):452–475
Ioris AAR (2017b) Agribusiness and the neoliberal food system in Brazil: frontiers and fissures of agro-neoliberalism. Routledge, London
Jepson W (2006) Producing a modern agricultural frontier: firms and cooperatives in Eastern Mato Grosso, Brazil. Econ Geogr 82(3):289–316
Li TM (2014) Land's end: capitalist relations on an indigenous frontier. Duke University Press, Durham and London
Little P (2001) Amazonia. Baltimore, John Hopkins
Lucidio JAB (2013) 'A Ocidente do Imenso Brasil': As Conquistas dos Rios Paraguai e Guaporé (1680–1750). PhD thesis, Universidade Nova de Lisboa, Lisbon
Luxemburg R (1951[1913]) The accumulation of capital (trans: Schwarzschild A). Routledge and Kegan Paul, London
Marston SA (2000) The social construction of scale. Prog Hum Geogr 24(2):219–242
Marx K (1976[1867]) Capital, vol 1 (trans: Fowkes B). Penguin, London
Marx K, Engels F (1970[1846]) The German ideology. Lawrence and Wishart, London
Massey D (2006) Space, time and political responsibility in the midst of global inequality. Erdkunde 60(2):89–95
Mészáros I (1995) Beyond capital: toward a theory of transition. Merlin Press, London
Mikesell MW (1960) Comparative studies in frontier history. Ann Assoc Am Geogr 50(1):62–74
Moreno G (2007) Terra e Poder em Mato Grosso: Política e Mecanismos de Burla 1892–1992. Cuiabá, Entrelinhas
Moore JW (2015) Capitalism in the web of life: ecology and the accumulation of capital. Verso, London and New York
O'Connor J (1998) Natural causes: essays in ecological marxism. The Guilford Press, New York and London
O'Flaherty R (1846) A chorographical description of West or H-Iar Connaught: written A.D. 1684. For the Irish Archaeological Society, Dublin
Pacheco de Oliveira J (2016) O Nascimento do Brasil e outros Ensaios: "Pacificação", Regime Tutelar e Formação de Alteridades. Contra Capa, Rio de Janeiro
Pichón FJ (1997) Colonist land-allocation decisions, land use, and deforestation in the Ecuadorian Amazon frontier. Econ Dev Cult Chang 45(4):707–744
Prado Jr. C (1967) The colonial background of modern Brazil (trans: Macedo S). University of California Press, Berkley and Los Angeles
Pred A, Watts MJ (1992) Reworking modernity: capitalisms and symbolic discontent. Rutgers University Press, New Brunswick, NJ
Rancière J (2009) Moments Politiques. La Fabrique, Paris

Ribeiro O (1963) Portugal o Mediterrâneo e o Atlântico: Esboço de Relações Geográficas, 2nd edn. Lisbon, Sá da Costa Editora
Richards JF (2006) The unending frontier: an environmental history of the modern world. University of California Press, Berkeley
Santos BS (1993) Modernidade, Identidade e a Cultura da Fronteira. Revista Crítica de Ciências Sociais 38:11–39
Sato J (2000) People in between: conversion and conservation of forest lands in Thailand. Dev Chang 31:155–177
Schwantes N (1989) Uma Cruz em Terranova. São Paulo, Scritta
Serrão JV (1980) História de Portugal: O Século de Ouro (1495–1580), 2nd edn. Lisbon, Editorial Verbo
Smith G (1989) Livelihood and resistance: peasants and the politics of land in Peru. University of California Press, Berkeley
Sodré NW (1964) Formação Histórica do Brasil, 3rd edn. São Paulo, Brasiliense
Solimano A (2017) Global capitalism at disarray: inequality, debt and austerity. Oxford University Press, Oxford
Trotsky L (1932) History of the Russian revolution (trans: Eastman M) Haymarket, Chicago
Turner FJ (1920) The frontier in American history. Holt, New York
Velho OG (2009[1976]) Capitalismo Autoritário e Campesinato: Um Estudo Comparativo a Partir da Fronteira em Movimento. Centro Edelstein de Pesquisas Sociais, Rio de Janeiro
Watts MJ (1992) Space for everything (a commentary). Cult Anthropol 7(1):115–129
Watts MJ (2018) Frontiers: authority, precarity, and insurgency at the edge of the state. World Dev 101:477–488
Webb WP (1952) The great frontier. University of Texas Press, Austin
Wilcox RW (2017) Cattle in the backlands: Mato Grosso and the evolution of ranching in the Brazilian tropics. University of Texas Press, Austin

Chapter 3
Placing the Agricultural Frontier of Mato Grosso, Brazil

Abstract The intricacies of one of the most relevant agribusiness frontiers in the world today—the north of the State of Mato Grosso, in the southern section of the Amazon, Brazil—are considered through a critical examination of place making. Vast areas of Amazon rainforest and Savannah vegetation were converted there, since the 1970s, into places of intensive soybean farming, basically to fulfil exogenous demands for land and agriculture production. That goes beyond the configuration of new places at the agricultural frontier, and starts with a qualitative intellectual jump: from place making on the frontier to place making as an ontological frontier in itself. It means that, instead of merely studying the frontier as a constellation of interconnected places, we examine the politicised genesis of the emerging places and their trajectory under fierce socioecological disputes. The consideration of almost five decades of intense historic-geographical change reveals an intriguing dialectics of displacement (of previous socioecological systems, particularly affecting squatters and indigenous groups, in order to create opportunities for migrants and companies), replacement (of the majority of disadvantaged farmers and poor migrants, leading to land concentration, widespread financialisation and the decisive influence of transnational corporations) and misplacement (which is the synthesis of displacement and replacement, demonstrated by mounting risks and a pervasive sentiment of maelstrom). Overall, there was nothing inevitable in the process of rural and regional development at the frontier, but the problems, conflicts and injustices that characterise its turbulent geographical trajectory were all more or less visible from the outset.

Keywords Mato Grosso · Agribusiness · Place making · Soybean · Future scenarios

3.1 Mato Grosso's Agricultural Frontier

The chapter will expand the conceptual and empirical analysis initiated in the preceding pages and will offer a further critical reflection upon socio-spatial trends and tensions related to frontier making. Following the framework introduced in

A. A. R. Ioris, *Frontier Making in the Amazon*, Key Challenges in Geography,
https://doi.org/10.1007/978-3-030-38524-8_3

Chap. 2—basically, the argument that socio-spatial frontiers are the result of simultaneous and synergistic production of abundances and scarcities in both established and newly incorporated areas (referred to as the law of scarcity–abundance)—the focus will be now on the range of interrelated processes that concur to shape the multiplicity of places at the frontier. It will particularly examine the vast areas of Amazon rainforest and Savannah vegetation in the state of Mato Grosso—in the centre-north of Brazil and located in the geographical core of South America—that have been converted, since the 1970s, into intensive farmland and growing urban settlements. Mato Grosso constitutes one of the world's fastest expanding hotspots of agribusiness activity and crop export (Deininger and Byerlee 2011), considered the most dynamic agricultural frontier in the Amazon today, where the successes and failures of agroecological management serve as an emblematic example of the trajectory of commodity-driven tropical agriculture (Brando et al. 2013). Our starting point here is that the semiotic and material frictions of modern agribusiness frontiers in the Amazon cannot be properly understood without reference to place making; that is, 'the set of social, political and material processes by which people iteratively create and recreate the experienced geographies in which they live' (Pierce et al. 2011: 54). In effect, agribusiness production in Mato Grosso has been a powerful place making engine, given that in just a few decades, it was transformed from a remote, largely forgotten part of Brazil into one of the strategic hubs of production and export in the country.

Especially in the Teles Pires river basin—situated in the north of Mato Grosso, right at the transition from Savannah to forest ecosystems—the landscape in the rainy season is dominated by the green colour of huge plantation farms. The intensification of agribusiness has created both sectoral relations where capital dynamically circulates (even if most of it leaves the region after the harvest) and large-scale socio-spatial interactions across multiple scales and successive agriculture cycles. The region is mostly occupied by first- or second-generation farmers, rural workers and commercial partners who now largely depend on the productivity of soybean and on its price in globalised markets. Those are all integral components of a dynamic, but unequal, place making phenomenon. However, the modernisation of agribusiness in relation to place making has not been properly recognised by scholars working on socio-spatial frontiers or on the political-economy of agri-food networks. On the contrary, the geographies of food and rural development have remained largely elusive regarding the production of space under globalising pressures and the resulting repercussions in terms of nested, place-based contradictions (Ioris 2016). Where the juncture between globalised forces and localised spatial outcomes has been acknowledged by social scientists, there is often limited conceptualisation and limited critical assessments of the place-related intersections between the intensification of agri-food systems and the complexification of specific time–space interactions (at farms, regions, countries and beyond), especially in relation to frontier making.

For most academics, agriculture happens in places that are already established and they normally focus on topics such as natural resources, techno-economic trends and socio-economic institutions (e.g. Duram and Oberholtzer 2010). By the same token, the conversion of eastern and southern tracts of the Amazon region into socio-economic frontiers is, to some extent, well documented (e.g. Foweraker 1981; Ianni 1981; Moran 1981; Sayago et al. 2004), but the contested and capricious place making associated with the agribusiness frontier has also received limited attention and narrow theorisation (D'Incao and Silveira 2009). In the Amazon region, a lot is known about the place making processes regarding the Transamazon motorway (Schmink and Wood 1992), about smallholder livelihoods and meanings (Vadjunec 2011) and land-related violence (Walker et al. 2011), but much less is known about industrialised agribusiness and its urban and regional repercussions in terms of social change and the advance of socio-spatial frontiers.

There is still a need for dedicated examinations of the intensely politicised processes of inclusion and exclusion mediated by the appropriation of the material and immaterial components of the lived reality of the socio-spatial frontier. In the case of the Teles Pires, a region previously characterised by exuberant natural scenery and numerous indigenous groups, has been irreversibly jolted by roads, new towns and (mostly) soybean fields. The transformation of the Teles Pires happened in less than five decades, always resorting to the monochromatic, but powerful, excuse of economic growth at any price. There was a constant promise of rationality, progress and welfare underpinning public policies and government action. To justify the imposition of the new agricultural frontier, the official discourse emphasised that it was 'no man's land', an empty space ready to accept the displacement of existing socionatural processes and their replacement, even if against the wishes of those already living there. Against what were considered rudimentary, excessively simplistic places, an even greater simplification was imposed: develop or die. Such extraordinary geographical trends challenge conventional analytical approaches and call for novel interpretative procedures. The next pages will help to fill that intellectual gap and demonstrate that the synthesis of this vicious dialectics between displacement–replacement is the resulting pervasiveness of misplacement in this agribusiness frontier of increasing global relevance. This reflexive examination is focused on the most emblematic municipalities in the Teles Pires—namely, Sorriso, Sinop and Lucas do Rio Verde—which were established after the construction of the BR-163 motorway, with a length of 1,777 km, to connect Cuiabá, in the state of Mato Grosso, with Santarém in the neighbouring state of Pará.[1] Before that, it is necessary to review the dedicated literature on place making and then shed light on key concepts, which are specifically relevant to this chapter.

[1]BR-163 in Mato Grosso was constructed by the Brazilian army (the ninth Battalion of Engineering and Construction); the work began in 1970 and it was inaugurated in 1976, but was strongly opposed by national and international environmental and pro-indigenous organisations. In spite of public investments and privatisation in recent years, a significant extent of BR-163 is still unpaved.

3.2 Place-Related Literature and Appropriate Analytical Sensibilities

Place making is a heuristic analytical category directly associated with the wider process of frontier making. It helps not only to understand the lived space, but also to uncover the discursive and contextual aspects of the agricultural frontier established in Mato Grosso. The attention given to place-based controversies complements and enriches the wider trend of interlocked abundances and scarcities. New areas of agri-food production, such as the Teles Pires river basin, are a locus of intense spatial reworking and relentless experimentation, which stretches from the specific and actually existing to the general and intertemporal. The agricultural frontier is more than just a zone of capitalist transition (cf. Barney 2009); it is a true 'laboratory' or 'smelting' of space with competing tempos and rival spatial rhythms. Because of its intense activity, agribusiness at the frontier of economic development is an accelerated mechanism of place making that is reliant on the rapid conversion of abundant territorial resources and on the promise of a better life. The resulting places are still relatively young, in terms of their politico-economic and social–ecological records of accomplishment, and their ontological novelty encapsulates emerging socioecological relations at the intersection between powerful global markets, regional development, widespread disruptions and interpersonal connections. However, it is not enough to meticulously and critically examine the places that form the new agricultural frontier. What is also needed, as an important element of our analytical approach, is a qualitative intellectual jump from place making on the frontier *to place making as an ontological frontier in itself*. It means that, instead of merely studying the frontier as a constellation of interconnected places, it is important to scrutinise the politicised genesis of the emerging places and their trajectory under socioecological disputes.

The nuanced ontological relationships between scales, times and dimensions are normally missed in conventional geographical texts, in which places are simply associated with the local and particular, in contrast with large-scale connections typically related to space. In the tradition of Vidal de la Blache, Fernand Braudel and others, place is taken as a slice, or a portion, of space with specific geographical qualities that are hard to generalise. Schematically speaking, place is unique, while space is more general. Such a rigid space–place dichotomy (which replicates other fruitless dichotomies, such as general–particular, anecdotal–historical and nature-society) certainly offers little help regarding the geography of the agribusiness frontier. Fortunately, we can find support in three alternative approaches that, although not necessarily applied to agricultural areas, lay emphasis on the social practices involved in place making and on the fluid connections between locales, places and other spatial scales (Agnew 2011). First of all, feminist scholars have pointed out that the identities of places are necessarily unfixed, contested and multiple, whereas local experiences are not only localised. Mundane, taken for granted activities, most practiced by women, also play a very important role in the making of places, which happens at the interconnection with wider politico-economic and globalised processes (Dyck 2005). That is particularly the case because domestic and caring labour are essential for the reproduction and stabilisation of capitalist relations of production (Mcdowell 2015). What gives places their specificity is the fact that they are constructed out of myriad social and physical relations spreading through networks and different scales of interaction. While there is a global and globalised sense of places and place making, the 'global is in the local in the very process of the formation of the local' (Massey 1994: 120). Consequently, grand narratives about place, and any other spatial formation, exert a highly negative influence, because they obscure the importance of gender, race, caste and age differences and, as a result, lower the possibility for genuine political change (Massey 1993). Instead of that, people's experiences and the zooming in on individual lives allows for a close look at the particularity and everydayness of place making and of being-in-place (Lems 2016).

In addition, humanist geographers, also in opposition to pedestrian interpretations, insist that senses and opinions are likewise active agents of place making. A place is not determined in advance, but is simultaneously performed and represented. Humanist geography underscores social difference, avoiding generalisations and recognising the centrality of place and its multiple forms (Arenas and Geisse 2004). Tuan (1976) emphasises the importance of understanding people's behaviour, conditions, ideas and feelings in relation to the intricacies of space and place. Such agent-based interpretations are directly concerned with questions about consciousness, experience and intentionality. Places are connected to the core of people's existence, their practices, feelings and social relations (Bartos 2013). They have both an intimate dimension and a socialising logic (Antonsich 2010). Intentionality is, thus, an emergent relation with the world and not a pregiven condition of experience (Ash and Simpson 2016); by the same token, subjective views of place and space influence conscious or unconscious acts (Schutz 1972). Place making is necessarily seen here as a relational phenomenon embedded in the activity of politicised, non-territorialised networks in which places are framed (Pierce et al. 2011) and places should be treated relationally in order to better cope with a series of interactions that run 'into' and 'out from' places, such as climate change, labour movements and trade relations (Darling 2009), but still maintaining a critical engagement with the sociopolitical ordering of the world, including social, corporate and government organisations (May 1996). Such humanist, and closely related phenomenological, explanations are often intertwined with the arguments of cultural geographers, who contend that the meaning and experience of places also involves cultural questions resulting from a cumulative spiral of signification (Thrift 1983). Individuals have multiple attachments to place and experience it according to their cultural, political and historical circumstances (Pocock 1981). Places are multifaceted cultural constructions, just as culture is the medium through which socio-spatial changes are experienced, contested and constituted (Cosgrove and Jackson 1987).

Another (third) important voice in this debate about the ontological foundations of place making has come from radical geographers underscoring the damaging repercussions of class struggle and related forms of socioecological exploitation that are place-based. According to this group of authors, class struggle is directly associated with the politics of place making and with processes of place homogenisation and differentiation. Harvey (2006) points out that places and other spatial arrangements, such as landscape and territory, assume specific attributes under the capitalist mode of production, given that the control of place is essential to secure command over material commodity exchanges and authority over the workforce (for instance, with the threat of moving jobs to other locations). Uneven development, which is a central feature of the capitalist economy, is to a great extent a product of place-specific outcomes of the process of class struggle (Das 2012). The complexity and variegatedness of place making in a capitalist society thus requires conceptions of abstract, structural power complemented by the politics of everyday life, which should account for interpersonal relations, networks of subjectivity and socio-economic systems (Bridge 1997; Ioris 2013). On the one hand, the contradictions between market rules and social relations around place are typically suppressed by authoritarian forces that try to minimise and forcefully repress opposition and discontent (Thornton 2010). On the other, the politicisation of place, as much as the dynamics of social change and the possibilities of political emancipation, are important elements of critical scholarship (Swyngedouw 1999). From this radical geography perspective, disputes over statecraft under capitalism are also shaped by the politics of place and related to the possibilities for political practice by creating particular terrains of conflict, determining the terms of access and influencing subjective experiences of political life (Chouinard 1990).

The convergence of these three streams of geographical thought—namely, feminist, humanist and class-based—can surely inform our conceptual framework for the examination of place making at the agribusiness frontier. Moreover, although place is connected to the construction of political thinking and the production of knowledge, questions about place making cannot be resolved at the level of theory only (Dirlik 1999). The frontiers of intensive agriculture expansion (agribusiness) in the world today are less about business (as a purely economic result) and even less about agro (as food), because the most central interest is the immanent existence of the frontier itself (as a capitalist world order in the making). The agribusiness frontier constitutes a highly idiosyncratic 'space–time envelope' (cf. Massey 1994) in which the being and the becoming are still actively struggling. In the case of the Teles Pires, the frontier was the result of government plans, accumulated socio-economic demands for land (in other parts of the country) and the search for new money-making opportunities. New places were created through a curious unfolding of forthcoming and opposing forces, which combined the drive to conquest and the impetus to abandon what was already sufficiently explored. While the frontier has produced novel social actors, it has been nurtured by the rhetorical fabrication of an outside world inhabited by those who can be friends but are more often seen as foes (i.e. its ontological basis also depends on the interference of others within and outside, cf. Qian et al. 2012). For all these reasons, place making at the frontier is not just the manufacturing of new spatial arrangements; it also opens up the possibility to better interpret places as ontological frontiers.

In addition to points highlighted above, the uniqueness of the agribusiness frontier in the southern Amazon requires the help of bespoke intellectual devices capable of scrutinising the multiple mediations, suspended certainties and disputed rationales related to the production of places. Luckily, the examination of the intricate ontological questions related to place making in Mato Grosso can find unexpected assistance in the ideas of a 'very special geographer': the recently deceased poet Manoel de Barros (1916–2014), regarded by the critics as one of the best authors of contemporary Brazilian literature. The need to liberate the reality from its prearranged place-based configurations, and recreate the world, permeates Manoel's long and incredibly original artistic construction. Manoel left a vast artistic production full of incredible images and lavish verses that basically deal with what is considered secondary or irrelevant (e.g. encounters with stones, birds, insects, horses, organic and decaying matter, the habits of scattered rural families, etc.). From the micro and insignificant, the poet constructs an argument about some of the most universal and unending questions of human existence. He understood that 'Things don't want to be seen by reasonable people' [*As coisas não querem ser vistas por pessoas razoáveis*] (Barros 2013: 278) and also that 'That which goes nowhere has a great importance' [*As coisas que não levam a nada têm grande importância*] (Barros 2013: 135). It is a fortunate coincidence that Manoel was born in Mato Grosso and spent his early years on the shores of the Paraguay River, where distances were immense and time seemed to move very slowly. In his words, his family lived 'in a place where there was nothing (…) and we had to invent' the world; 'invention was required to enlarge the world' and 'disturb' the existing, normal meaning of things.

Manoel's fundamental ontological proposition was that Mato Grosso had yet to be 'invented' in order to decipher still unarticulated truths. Manoel realised, since his childhood, that the immensity of Mato Grosso was incomplete and, consequently, his world had still to be created, that is, the intense and sophisticated exchanges with nature and the small number of inhabitants needed to be complemented with broader social intercourses and connections with wider Brazilian and international society. A new reality needed to be invented was not only to be true, but because it was necessary to unlock the deep structures of the existing world. Central to Manoel's ontology is the difference between 'invention' and 'lie', in other words, the realisation that invention is diametrically in opposition to falsehood. In what was probably the only public interview ever given by Manoel de Barros—turned into the documentary 'Only Ten Percent is Lying' [*Só Dez por Cento é Mentira*] by Pedro Cezar, released in 2010—[2] Manoel claims that only 10% of his argument is untrue and 90% is invented [*Tenho uma confissão a fazer: noventa por cento do que escrevo é invenção. Só dez por cento é mentira*]. In his verses, he already stated that 'all that I didn't invent is false' [*Tudo que não invento é falso*] (Barros 2013: 319). However, the new reality needs to maintain the organic ontology that rightly encompasses everything, including organisms, people, stones, fluids, landscapes and unsaid sensations. In his highly original poetry, 'The trees commence me' [*As árvores me começam*] (Barros 2013: 311). Inspired by Manoel's provocation about the world

[2]Information about the documentary can be seen at: www.sodez.com.br. The poet was extremely reserved and his work was only revealed to the wider public in recent decades when some well-known intellectuals started to praise the relevance of Manoel's poetry.

to be invented, we will demonstrate the troubling trajectory of places at the Teles Pires frontier and how place making there has unfolded through a disconcerting dialectics of displacement–replacement–misplacement.

3.3 Frontier Making Through Displacement (1960s-Early 1980s)

In stark contrast to the supposed spirit of 'democracy and egalitarianism' famously associated with the American West by Frederick Jackson Turner,[3] the agricultural frontier in Mato Grosso was initially based on a widespread practice of displacement. For many generations of Brazilians, the north of Mato Grosso was a faraway place, a universe apart and homeland mainly to secluded first nation peoples (Buarque de Holanda 1994). That began to change in the first decades of the twentieth century due to ideological calls for modernisation, progress and integration of areas considered wasteland. The federal administration launched the March towards the West in 1937 and the Roncador-Xingu expedition in 1943 to fill the geographical voids still uncomfortably visible on national maps (Villas Bôas and Villas Bôas 1994). But the integrationist project didn't include the locals and their socioecology; on the contrary, the rest of the country fell over the region bringing back the spectres of Alexander, Alaric and Pizarro. New places started to be forged out of the remnants of the cultures, values and cultures of those who used to live in the region (indigenous groups and squatter peasants) and also out of the destruction of socioecological communities. The occupation of the north of Mato Grosso, financed and stimulated by state agencies, happened through the widespread and systematic grabbing [*grilagem*] of indigenous land, which amounted to an 'authentic ethnocide and genocide' of many tribes (Oliveira 2005: 84).[4] In tandem with initiatives undertaken by the national government, the state (i.e. provincial) administration systematically sold large tracts of land at very low cost to property speculators, certainly without much interest in exploring the areas. Vast state-owned areas, with hundreds of thousands of hectares—described in Portuguese as *glebas*—were easily transferred to new owners by corrupt officials, who typically expelled the local residents, used the areas as collateral for bank loans or resold them to colonisation firms (Moreno 2007). The fact that these *glebas* were demarcated from distant offices without any fieldwork opened the door to major imprecision, inadequate property boundaries and monumental fraud.

The decisive phase of spatial transformation came with the resolve of the ruling military between 1964 and 1985 to force agriculture development upon the remote

[3]Obviously democracy and development only applied to white settlers.

[4]See Asselin (1982) for a detailed description of how *grilagem* works and widespread corruption underpinning the operation of government agencies. Interestingly, the Brazilian case is not too dissimilar from the comparable experience in the United States where, according to Churchill (1997: 290) one third of the land of the 48 contiguous states 'possesses no legal basis for its use and occupancy'; North American '*grilagem*' has taken place mainly in the territory of indigenous groups, leading to the exploitation of minerals, including the very strategic mining of uranium.

corners of the Central-West region. Three National Integration Plans and other similar programmes, with international funding, were introduced in the 1970s. The state of Mato Grosso was actually considered the 'paradise of private colonisation' projects (Oliveira 1989: 106), which since 1974 replaced the initial focus on public farming schemes (Santos 1993). It was essentially a counter agrarian reform process that played a crucial role in the spatial expansion of capitalism in the country (Ianni 1981). Impoverished small farmers were brought mainly from the southern states to try the same strategy adopted by their ancestors, who previously had to leave Germany and Italy and moved to Brazil in search of a piece of land and a secure future (Schwantes 1989). As it is widely known, social mobility was notoriously restricted in the Brazilian countryside, but the frontier raised the promise of social betterment and the possibility to own a much larger property. Early research conducted by Oliveira (1983) ascertained that families of farmers coming from the south were sincerely in search of better conditions in the context of the progress announced by the government. In practice, poor peasants and small farmers struggled to reinitiate their lives in the adverse places of the frontier. The situation was worse for the indigenous populations, who could either move to precarious and fragile reservations or be decimated by diseases and abject exploitation. Besides the activity of individual farmers, private companies were encouraged to acquire land in the frontier with the promise of generous public incentives and subsidies, although these were often siphoned off to finance activities and enrich people in other parts of Brazil (Cardoso and Müller 1977).

The overall course of events was similar, but in the opposite direction, to the genesis of capitalist farming described by Marx. If in Europe, 'the expropriation of the agricultural producer, of the peasant, from the soil is the basis of the whole process' and the pillar 'of the capitalist mode of production' (Marx 1976: 876, 934), what happened in Brazil was the displacement of peasants from their original places followed by a poverty-induced migration to Mato Grosso that was again based on the displacement of the existing socio-spatial configurations. Those internally coupled processes of displacement (i.e. in the south and in Mato Grosso) happened as a coordinated form of primitive accumulation imposed by the political centres of Brazil, that made pervasive use of fraud and violence. For instance, the town of Sorriso, in the centre of the Teles Pires region, was a result of the occupation in 1972 by private developers of a property that belonged to the farmer Edmund Zanini, who was basically deceived and eventually displaced from the area he had purchased in 1964 (Folha de São Paulo 2009).[5] Not by chance, in a book by the local historians Dias and Bortoncello (2003), the brutality of displacement is concealed and Sorriso is praised as a beacon of abundance and economic growth that compensated for the 'loss of the paradise' in the south of Brazil (this is epitomised in the words of the poem 'My Place' on the back cover of the book). Three decades later, Sorriso, located at the edge of the forest and with huge areas of easily cultivable land, is now the main hub of soybean production for the entire country (Jepson et al. 2010).

[5]The Zanini family fled Mato Grosso in 1977 and the dispute was only resolved by the courts in 2011.

The town of Sinop, which is now the most important administrative centre in the region, had a similar trajectory of settlement associated with displacement. The process started with the acquisition of 645,000 hectares in 1970 by a colonisation company (itself called Sinop), a property known as Gleba Celeste. The urban area of Sinop started to be opened in 1972 and soon had more than 500 timber mills processing the result of intensive deforestation. The head of the colonisation company, Ênio Pipino, famously declared that Gleba Celeste 'was a green world, sleeping, in the loneliness of the Amazon' (in Souza 2006: 144) and also that he was 'planting civilisations' and creating a liveable Amazon by opening roads and clearing forests and jungles (Pipino 1982). The vector of displacement underpinning place making continued unabated, as a persistent phenomenon based upon dispossession and constant movement rather than upon stability, through which different layers of belonging, ties to land and group identity could be revealed (Connor 2012).

Despite the immediate material success of the new agricultural frontier (in terms of settling people and removing the original vegetation), at the same time as new farmers continued to arrive in the Teles Pires region, most ruined entrepreneurs left for other parts of the Amazon and beyond. Tragically, a significant proportion of those who came in search of their own piece of land eventually returned to their places of origin in the south or, alternatively, had to find employment in the increasingly large agribusiness farms (Barrozo 2010). This was particularly evident in the municipality of Lucas do Rio Verde, which relocated 203 families of small farmers from the state of Rio Grande do Sul (Oliveira 2005); after a difficult beginning, just a minority of the original pioneers remained in Lucas do Rio Verde—less than 10%—while most lost their properties due to the operational adversities, unfulfilled promises and accumulated debts (Oliveira 1989; Santos 1993). The fact that the frontier was strategically open for just a relatively short period of time in the 1970s and 1980s suggests that the vector of displacement conceded some space to its opposite—replacement—in order to consolidate the meaning of the emerging socio-economic and politico-spatial relations.

3.4 The Inventive Force of Replacement (End of 1980–2000s)

The previous section discussed how the making of new places at the frontier was achieved through displacement, (involving the arrival of thousands of migrants in the short interval of only a few decades), the large-scale removal of the original vegetation, and the introduction of a gradually more intense production. The mechanics of displacement was hegemonic in the attempt to overcome the residues of pre-capitalist society, impose a new spatial order and facilitate the access to territorialised resources (land, water, timber, bushmeat, etc.). However, displacement could not happen in isolation and, even as the existing socio-spatial features were being displaced, another key force—replacement—was emerging. Especially from the late 1980s, some of the groups initially attracted to the agribusiness frontier were becoming redundant and had to swiftly adapt to a reality fraught with unexpected difficulties. It became increasingly evident that land in the new places had never been available to all newcomers. Less productive workers and decapitalised farmers lost their position and were largely replaced by a small number of skilled machine operators (trained to cope with the rapid automatisation and informatisation of farming procedures). The initial movement of attraction and displacement was followed by an increasingly strong process of repulsion and replacement.

In the first period of the agribusiness frontier, due to the political and economic risks involved, a large number of migrants were essential for the consolidation of the frontier, thus justifying investments in infrastructure, securing policy concessions and satisfying public opinion that something was being done about agrarian tensions in the south. Soon after this foundational moment, several political and economic constraints affected the ability of the federal administration to keep the doors wide open (in particular because of a public debt out of control and high rates of inflation in the late 1980s). In this way, the same frontier that attracted migrants, as a seductive mirage and promise of a better life, began to expel a significant proportion of those who moved to Mato Grosso. For such vulnerable players, there were basically three main options: become labourers in rural properties, try to receive a small plot in agrarian reform projects or transfer their activity to a small farmstead near to the towns.[6] As observed by Marx (1976: 905), the capitalist farmer results from the enrichment of some individuals who usurped the common land (with the impoverishment of 'the mass of the agriculture folks') and managed to benefit from technological revolutions. Land concentration was fuelled by significantly higher land prices following the economic success of the frontier, which in Mato Grosso increased by 514.1% between 2002 and 2013 (compared to the national average of 308.1% in the same period), according to an assessment by the Ministry of Agriculture and the University of Brasília published in 2015.

Replacement was not only restricted to the concentration of landed property and the conversion of the weaker farmers into farm labourers. It involved other profound changes in economic and technological trends, including the substitution of the various crops unsuccessfully tried in the 1970s (coffee, cassava, guarana, pepper, rice, etc.) with the overpowering presence and symbolic importance of soybean (predicated on the use of intense agronomic techniques, expensive machinery and financialisation of production). In this particular context of place making, the soybean was victorious from the outset, inevitably, because it played a central role in the consolidation of a model of regional development reliant on crop exports, in the hands of large farmers and transnational corporations. The Teles Pires has in effect become a large soyscape and those who controlled soybean controlled the flows of money. Since then, what really started to matter in the region, and affect the social

[6]To avoid the replacement of those who had just been replaced, municipal authorities introduced an informal 'place filter' that prevented the entrance of poor, unwanted migrants: for a while, at the bus station of Sorriso and Sinop there was a formal check and those unable to demonstrate secured income receive a free ticket back home.

status of most people, was the overpowering entity 'soybean-money'. The production of soybean almost doubled every year during the 1980s in Mato Grosso and continued to grow in double figures in the following years. The farmers also became better organised and created, in 1993, a highly effective technological institute, the MT Foundation, responsible for the development of new soybean varieties that are more productive and disease resistant. Yet the advance of soybean production was not linear. Because of state reforms and monetary stabilisation plans, the early 1990s constituted a challenging period for the Brazilian agriculture sector. Agriculture was increasingly influenced by events taking place outside the sector, including trade liberalisation, deregulation, credit reforms and removal of price support policies.

After a moment of great turbulence, there was a revitalisation of the frontier since the end of the decade, helped by currency devaluation in 1999, foreign investments in productive, and speculative, ventures; and growing demands from Asia (especially from China). Many transnational corporations (TNCs) were attracted to the Teles Pires in the period between 1999 and 2005, when booming commodity prices resulted in a sizable increase in crop production under the influence of replacement pressures. There are also other political and symbolic repercussions of the uncompromising replacement of farmers and technologies in the Teles Pires. Soybean production has been constantly portrayed by sector representatives as a fine expression of technological efficiency and administrative know-how, which is used as undisputed evidence that rational, high-tech development works. This strong defence of soybean agribusiness was also helped by the fact that rural leaders increasingly entered the politico-electoral system (such as mayor Pivetta, vice-governor Fávero and, most famously, former governor, senator and secretary of state for agriculture (between 2016 and 2018) Blairo Maggi, owner of one of the most powerful Brazilian TNCs). The prevailing claim is that technified agribusiness has replaced the tradition of chaos, incompetence and turbulence typically associated with previous rounds of economic development in the Amazon with a new socio-spatial reality based on rationalism, knowledge and competence. It is an essentialist perspective by those who control place making that, in practice, constantly denies alternative forms of agriculture or a different socio-economy. The symbolism and rhetoric of the successful frontier plays an important role in the definition of the new agriculture places against other possibilities who are outside (what Massey 1994, describes as the production of selective inclusion and also of boundaries of exclusion).

However, claims of success and technological innovation are insufficient to conceal the mounting contradictions of the agribusiness frontier. According to the federal environmental agency IBAMA, soybean accounted for most of Mato Grosso's environmental problems because of deforestation (Repórter Brasil 2010), while the state was responsible for 70% of national deforestation in the year 2015 (Garcia 2015). Figure 3.1 illustrates recent deforestation in the Teles Pires River Basin in 2015. As in most of the Amazon region, agribusiness development superimposed an urban logic, and globalisation tendencies, over regional place making (Rempel 2014). Less than 30% of the population now live in the countryside and landowners typically live in the cities and commute every day, only spending more time in the rural property during seeding and harvesting periods. Those towns are agribusiness municipalities

Fig. 3.1 Recent deforestation in Mato Grosso

with high levels of urbanisation and a range of specialised services to attend to the demands of modern agriculture (including logistics and financial services), but also with marked contrasts between the wealthy centre and a growing urban periphery consisting of low-paid workers and the unemployed (Elias 2007). There are sustained cases of racial and socio-economic discrimination against those in the periphery; normally those who came from the northeast or other parts of the Amazon and who are typically non-white migrants (such as the majority of the residents interviewed in the periphery of Sinop and Sorriso, respectively, in the deprived neighbourhoods of Boa Esperança and São Domingos). The mismatch between the positive image of the frontier and the crude experience on the ground produces a tough synthesis to pull off.

3.5 The Resulting Sense of Misplacement (Since the 2000s)

From the above, it can be argued that place making in the Teles Pires produced urban and rural landscapes of intense economic activity that are also fraught with difference, tensions and inequalities. The high-tech agriculture practiced in the Teles Pires did secure national and international prestige among agribusiness players and is now widely praised for its productivity, rationality and entrepreneurialism. At the same time, there are striking contrasts, for example, between wealthy urban areas and agribusiness farms on the one hand, and the poverty of urban peripheries (Fig. 3.2) and small family farms on the other. Because of its unique genesis and turbulent advance, it seems that there is more than just ostentation and socio-spatial inequality in the agribusiness frontiers of the Teles Pires. Despite signs of progress and opulence, place making in the Teles Pires continues to be in a state of great uncertainty and complex economic, technological and social constraints. One main source of instability is the fact that, because of the politico-economic crisis of the 1990s, the region was inserted too easily into the circuits of global agri-food markets and neoliberal economic reforms (Ioris 2015). Public and private life has been affected by those adjustments which, despite renovating the regional economy, reinforced the pattern of socioecological exploitation, vulnerability and political subordination to the main centres of power.

What is also particularly remarkable in the case of Mato Grosso is that the unsettling dialectics of displacement and replacement continues to define place making in the region long after the opening of the agricultural frontier. Present-day circumstances remain greatly based on the original mechanisms of territorial conquest and political control put in practice since the middle of the last century. Figure 3.3 shows a grain storage facility near Sorriso owned by a TNC. The violent displacement of the earlier socioecological condition was not followed by a condition of spatial stability, but was instead complemented, and magnified, by a never-ending replacement of

Fig. 3.2 Low-income periphery of Sinop

Fig. 3.3 Grain storage facility near Sorriso

people, knowledge and social practices. Rather than the more common succession of displacement by emplacement (as the consolidation of the spatial configuration that probably characterises most agricultural frontier areas), what happened in the Teles Pires was the perpetuation of displacement by new waves of replacement. It is precisely this synergy between displacement and replacement that facilitated the employment of some of the oldest methods used in the Brazilian countryside, such as the exploitation of the workers and large-scale deforestation. The agricultural frontier was established to serve, and continues to attend, primarily, the politico-economic agendas of such powerful economic groups in the region and, most importantly, outside Mato Grosso. The consequence is place making embedded in trans-spatial flows and international networks through power exercised extraterritorially.

All things considered, place making in the Teles Pires continues to move at a fast pace, but remains based on a fundamental paradox between the presumption of progress and collective achievement, and the concealment of the fact that most social and economic opportunities are increasingly restricted. While agribusiness is ubiquitous—not as merely an economic activity, but as the holy grail of modernisation—in reality, it is touched by very few (as in the case of large-scale irrigation in Fig. 3.4). As observed in an interview with an advisor of the agriculture federation FAMATO, 'there is a lot of technology available today, but it is in the hands of only a few, the big [farmers]'. The local population now lives a strange, increasingly troubled, disconnection between the proclaimed success of the agriculture frontier and the emerging

Fig. 3.4 Centre pivot irrigation near Lucas do Rio Verde

realisation that not everything corresponds to those claims. There is a rising concern with, among other issues, the long-term viability of soybean production; the risks of a very narrow economic base; the isolation of the region in relation to input suppliers and soybean buyers, and the hidden agenda of politicians and sector representatives.

These suggest that several decades of the spatial dialectics of displacement and replacement actually resulted in a pervasive, although often silent, sentiment of misplacement. Despite all the positive images transmitted daily in the local and national media, the region seems misplaced, its future is ambiguous and most of the population still struggle to reconcile being and belonging. New places have been produced, and afterwards many have been destroyed, because of the alleged advantages of the agriculture frontier, whereas these are, in effect, signs of great weakness. Moreover, misplacement is not a passive synthesis of displacement and replacement, but it is actually the third term of a highly idiosyncratic 'trialectics' (cf. Ioris 2012) and, therefore, has also become an active driving force in the process of place making. For instance, the sense of misplacement in the Teles Pires is appropriated by the hegemonic groups and then used as justification for new rounds of capital accumulation under strong calls for efficiency, better logistics and competitiveness. The fact that misplacement is the dialectical synthesis of the interplay between displacement and replacement reveals the full extent of the colonisation of space by capital and the production of abstract space, as long argued by Lefebvre.

Finally, the recognition that misplacement has also been converted into a force for place making has another unexpected and probably surprising result: the progressive and disturbing shrinking of space in the Teles Pires. In other words, the regional space has not only been produced through place making, but has also been wasted, corrupted and ultimately diminished. If the physical map of Mato Grosso retains the same nominal area (around 90 million hectares) and the Teles Pires has the same boundaries as 40 years ago (with an increasing number of municipal authorities), the social and socioecological space has gradually reduced year by year because of the perverse model of place making adopted to open the agribusiness frontier. The process of place making has relied on the acute degradation of nature, the destruction and waste of timber, land and biodiversity, and on the subjugation of those who came to the region naively in search of an improved future. An important element of the reduction of space when the agricultural frontier advances is the decoupling of intense farming from food production. This problem is not unique to the Teles Pires region, but it is particularly embarrassing that the main area of agribusiness production in the country is, in effect, a large food desert where most of the basic staple foods, such as rice, beans and vegetables, are imported from other Brazilian states. Likewise, the concentration of land (a typical agribusiness property has between 1,000 and 2,000 ha) led to an agriculture system that is less efficient per area of cultivation when compared with smaller properties, especially those below 100 ha (Helfand and Levine 2004). Taken as a whole, the prevailing direction of place making under the influence of agribusiness has produced a reality of prevalent misplacement, in which places are less ecologically viable, more unstable and even smaller than the previous socio-spatial situation before the opening of the frontier.

3.6 In the End, the Frustrated 'Invention' of Mato Grosso

Due to the convergence of developmentalist policies and the attraction of large contingents of migrants, the northern section of Mato Grosso has become one of the last, and most important, frontiers of agricultural expansion in the world. Instead of a gradual advance of private property and market transactions, the government planned and imposed new places upon vast areas and easily mechanisable tablelands in the Teles Pires since the early 1970s. The prevailing mechanisms of place making have been the propagation of displacement through a recurrent replacement of people and conditions of production, which results in a widespread sense of misplacement. This hegemonic direction of regional development followed a very different trajectory to that envisioned by Manoel de Barros for his homeland. In the 1920s, the poet wished for an 'invention' of Mato Grosso to replace a reality fraught with anachronisms and subject to spatial forces that isolated people into remote communities. Manoel's main proposition was to reconfigure those places and realise human potentialities at the same time ('Good is to fit in the landscapes as a river, a stone' [*Bom é constar das paisagens como um rio, uma pedra*], cf. Barros 2013: 390). The poet wanted a new spatial order that was more inclusive and less fragmented. Manoel also warned about a rival pathway, which was qualitatively inferior and would produce a misleading reality based on lies and wrongdoing. However, from the empirical evidence available now, there is plenty of material to infer that Manoel's stipulation was not upheld. On the contrary, the geographical typology provided by Manoel—that is, the difference between invention (as something genuine and positive) and falsehood (as inauthentic and dubious)—helps us to realise that place making in the Teles Pires has been an accumulation of lies, instead of the proper invention of the world. That happened through another crucial paradox (in a long sequence of perverse paradoxes, some discussed above): what was considered too simple a space was displaced and replaced with an even simpler space, which is only deceptively more sophisticated or more advanced.

What existed before had to be violently displaced through the firm hand of the state and the involvement of a large number of impoverished farmers from the south of Brazil (and also some business enterprises in search of the easy, subsidised government incentives). The region was opened up to public and private colonisation schemes and rent-seeking companies in an intense place making process boosted by the state through the construction of roads, airfields, storage facilities and the growing expansion of urban settlements. Soon after the frontier was considered irreversible, there was potentially an opportunity to accommodate the needs and aspirations of all those initially involved. Not even that was achieved. Although at first, the aim was to occupy areas considered (or made) empty and cope with major structural deficiencies in the best way possible; since the 1980s, the main driving force was to replace the promise of land for all and emphasise high-tech, efficient agribusiness production as the only way forward. Instead of making the world bigger, as Manoel wanted, place making has been characterised by spatial compression through the accumulation of land and accelerated financialisation of production (particularly under the sphere of influence of TNCs and private banks). The ultimate result is that Mato Grosso's space has been shrinking since the early days of the agricultural frontier, due to sociocultural and socioecological erosion.

Overall, the agribusiness frontier in the Teles Pires is not only a chain of numerous places that are profoundly interconnected, but the new places also reveal a great deal about tensions related to spatial change and are themselves geographical frontiers. Beyond the apparent uniformity of crop fields and the homogeneity of plantation farms, there are major social inequalities, the almost forgotten genocide suffered by indigenous groups and the risks of a socio-economy reliant on a single activity (soybean). Although the advocates of agribusiness make optimistic claims about the positive qualities of the new places, they systematically pursue strategies that are inherently partial and leave most of the population and socionature behind. That leads us to a final and very disconcerting observation: there was nothing inevitable in the process of rural and regional development promoted in the Teles Pires, but at the same time, the problems, conflicts and injustices that characterise its turbulent geographical trajectory were all more or less visible from the outset. Very little could have been different, considering the past process of territorial conquest in Brazil and the brutal advance over the Amazon in the previous century. In other words, the new places at the frontier have been impregnated with the worst forms of money making, aggression and racism. The consequence is that, more than the soybean, the deceptiveness of place making is the main contribution of this agricultural frontier to the rest of the world. In the final section, it will explore the most emblematic future scenarios of frontier making in the state of Mato Grosso.

3.7 Looking Forward: Future Scenarios of Frontier Making

It was examined above the evolution of agricultural frontier making in Mato Grosso (as a representative example of the modernisation of agriculture along the lines of the expansion of agribusiness to new areas in Brazil), which has been permeated by undeniable achievements in terms of the expansion of the areas under cultivation and the increasing circulation of capital. Under the influence of a neoliberalising platform, the sector is now one of the pillars of the national economy and the main driving force of the Mato Grosso economy, particularly because of the export of soybean to Asian countries and other international markets (Ioris 2017). At the same time, the success of agribusiness has been counterbalanced by the extension of social and ecological impacts, the violence associated with the consolidation of the agricultural frontier and the need to constantly justify the concentration of land and opportunities. Considering the evidence available and the reactions of those contacted during the research, it is possible to expect that the evolution of the agribusiness frontiers of Mato Grosso will either maintain its current trajectory of intensification and high profitability (possibly with the mitigation of the most evident impacts through the adoption of new technologies and the enforcement of the existing labour and environmental legislation), require more determined state interventions to mitigate mounting tensions, or that the sector will gradually decline with a reduction of productivity and production areas (possibly due to higher costs, phytosanitary risks or logistical difficulties).

It is obviously not possible to predict what will happen in the next decades, but the development of neoliberal agribusiness will certainly unfold according to the balance of power and other socioecological factors that simultaneously, and contradictorily, attempt to either promote or restraint crop production. For instance, the 2008 global financial crisis, which took more than a decade to be (circumstantially) resolved through a concerted effort that also involved the production of cheap commodities in global frontiers, demonstrated the fragility of market liberalism and the multiple inequalities permeating both moments of rapid economic growth and recession periods (Dardot and Laval 2010). Nonetheless, if no significant problems occur, and given the macroeconomic relevance of primary exports for the Brazilian economy, agribusiness activity tends to retain its decisive political and ideological position (particularly after the election of a right-wing government in 2018). In the 12 months to January 2019, agribusiness exports accounted to US$ 102.1 billion, the equivalent of 42.3% of total Brazilian exports and responsible for a surplus of

US$ 88.1 billion in the period, according to the statistics published by the National Confederation of Agriculture (CNA). However, at the same time, there are signs of a growing reaction and discernible sources of risks that can also impose major difficulties for the persistence of current trends.

The majority of our interviews, conducted in the last few years, revealed this central dilemma between the accomplishments and the limitations of agro-neoliberal trends in Mato Grosso. Some respondents emphasised the negative aspects, while others were clearly impressed by the economic and social transformation of the region. Taking into account such diversity of opinions (note that the interviews had specific questions about current conditions and possible future developments) and the characteristics of an economic sector that depends so heavily on the cultivation of a few crops (soybean, cotton and maize, in special), it was possible to summarise the future of agribusiness in Mato Grosso under three main scenarios with a time horizon of 8–16 years (over the next one or two presidential mandates, including re-elections after 4 years), considering key reinforcing processes and potential disrupting events (also in Table 3.1):

Table 3.1 Future scenarios of agro-neoliberalism in Mato Grosso (Brazil)

Scenario	Scenario description	Hypothesis regarding main trends	Hypothesis regarding disrupting events
Expansion and consolidation of agro-neoliberalism	Preference for market-based strategies (including public–private alliances)	✓ State apparatus with merely a supporting and regulatory role ✓ Heavy dependence of the regional economy on agribusiness	✓ Dramatic oscillation of market prices ✓ Politico-economic disruption
Return of strong government interventions	Recovery of state protagonism and direct state interventions	✓ Farming community requesting a return of direct state support ✓ Greater availability of public funds	✓ Inefficiency of state interventions in production activities ✓ Ideological and political pressure for market-based solutions
Containment and decline of agro-neoliberal trends	Reduction of cultivated areas and the influence of export-driven agriculture	✓ Phytosanitary and climatic risks ✓ Strong protests in favour of social reforms and agroecological options	✓ Space for authoritarian measures ✓ Lack of national and international support

3.7.1 Scenario 1—Expansion and Consolidation of Agro-neoliberalism

Because of the profitability and favourable prices in global markets, the production of soybean and other crops has continued to expand, though at lower rates than earlier when it reached 6% a.a. The sector managed to overcome the reduction of government funding and difficult market situations (as in the year 2005) through a successful mobilisation of political forces, strategic alliances with transnational and national corporations (which are increasingly responsible for financial support) and the adoption of new technologies (such as genetically modified soybean and more efficient machinery). The agribusiness industry of Mato Grosso has occupied an important space in the environmental debate and appropriated the rhetoric of sustainability—although through a highly technocratic and non-political angle—what has helped to improve its image nationally and internationally. The sector has been firmly inserted into global trade and financial transactions, including sizeable foreign capital flows to corporations in the region through tax havens, leading to reduced transparency and substantial losses of tax revenue (Galaz et al. 2018). In that context, instead of fighting the environmental regulators, the agribusiness sector has claimed to obey and to be a major ally of those concerned with ecological conservation. A stronger and less controversial agro-neoliberalism has echoed the calls for innovation and sustainable development put forward by Alegria (2005), which was considered an impetus towards greater collaboration between groups and organisations in more constructive and spirited ways. At the global level, it corresponded to the New Vision for Agriculture, defined by World Economic Forum partners in 2009, as technological and institutional adjustments sufficient to increase agriculture by 20% per decade until 2050; according to this perspective, the main strategy was the leveraging of market-based approaches through coordinated efforts by all stakeholders, including farmers, government, civil society and the private sector.

3.7.2 Scenario 2—Return of Strong Government Interventions

This second scenario is actually a variation to the previous one, given that the apparatus of the Brazilian state has never really left the stage. On the contrary, the state remained the main player of agriculture production and regional development, as well as the main promoter of neoliberalising policies, such as the integration with global markets, support to individual farmers and the privatisation of roads and infrastructure. However, it was possible to envisage that because of acute market turbulence, reduced appetite of transnational corporations for the region or strong political pressure by the agribusiness community, the state was again required to exercise more leadership and intervene more directly. As demonstrated in the global report published by McIntyre et al. (2009), the state could resume its developmentalist role if necessary to overcome difficult situations and to coordinate efforts of different groups and sectors. That was proved indispensable particularly when local or national political groups questioned the political settlement that underpins the growth of agribusiness in Mato Grosso. In any case, such periods with stronger state initiatives did not necessarily cancel the wider evolution of agro-neoliberalism, but represented a correction of excessed and transition to a new phase.

3.7.3 *Scenario 3—Containment and Decline of Agro-neoliberalism*

Contrasting with the previous scenarios, the internal contradictions and negative reactions to the elitist basis and uneven development of agro-neoliberalism in Mato Grosso (see Vieira et al. 2014) led to its weakening and even collapse. Similar disruptive situations had happened before in the environmental history of Brazil, as in the case of rubber, cocoa and (in recent years) orange production. There were many phytosanitary and environmental risks associated with the cultivation of large extension of land with a single crop and, in the case of soybean, almost a single variety of genetically modified plans. The biological vulnerability of monoculture was greatly aggravated by climatic changes that resulted in pronounced droughts, more intense rains and hotter growing seasons. In addition, there were unresolved and widespread conflicts in the main production areas of Mato Grosso related to land tenure, demands of indigenous tribes and uncertain conservation reserves. Those tensions re-emerged and intensified when new roads and hydropower schemes are proposed, which have been a regular occurrence (especially because of the attempt to reduce transportation costs through the improvement of roads and river navigation, as well as the construction of hydropower projects to respond to local and national electricity demands). Related to local disruptive events, national and international pressures for the reduction of deforestation for a more stringent environmental regulation and for agroecological alternatives contributed for the decline of the agribusiness appeal.

The above three main scenarios—as narratives or images of the future—were developed independently (that is, in full ignorance of the similar results of other authors), but these have important points of convergence with some of the scenarios developed by the Global Scenario Group (see Raskin et al. 2002) or comparable exercises, as the one conducted by Bourgeois (2016). Such international experts tried to envision possible directions of global change through a detailed assessment of key driving forces, quantifiable indicators, critical barriers and also desired outcomes (as in the case of a transition to higher levels of sustainability). The final report summarised their findings under three classes of scenarios, namely Conventional Worlds (no major changes in current market and policy trajectories), Barbarization (problems are not managed and crises are amplified) and Great Transitions (involving profound transformations of values and the organising principles of society). Those overarching classes are further expanded into six scenarios, of which three seem to have relevance to the interrogation of the future of agribusiness in Mato Grosso. One is the scenario 'Market Forces' which is optimistic about market-based solutions and is fundamentally related to the self-correcting logic of competitive markets; this is closely connected with Scenario 1 described above, that is, the continuous expansion of agro-neoliberalism. A second scenario put forward by the Global Scenario Group is 'Policy Reform' which entails stronger policy guidance and renewed government actions; this has correspondence with Scenario 2, which describes the return of direct interventions by the state apparatus.

Finally, the international report presents a future scenario described as 'Breakdown', basically the collapse between market forces and the systemic failure of government initiatives, leading to conflicts and fragmentation. This last scenario has disturbing parallel with Scenario 3, which projects the rapid decline of agribusiness in Mato Grosso due to mounting management contradictions and the barriers offered by nature. It is an aim of the scenario community to determine which scenario is more likely (Öborn et al. 2011), but perhaps it should be pointed out that, since the early days of conquest and exploitation, the Amazon region has been marked by violent transformations shaped by mercantile dynamics and the exploitation of its socioecological features. If the experience of the past four centuries can be used to speculate about the future, it may be possible to expect that in the next decades neoliberalised agribusiness in the southern Amazon region, as in Mato Grosso, would be compromised and perhaps collapse due to its intrinsic vulnerabilities and the impossibility to harmonise clashes and tensions through market or state mediation. This third, distressing scenario will essentially be the actualisation, through the advance and downfall of agro-neoliberalism, of this long-lasting tradition of violence against local peoples and their socionatural relations that have long marked the geography of the Amazon.

3.8 Conclusions: The Prospects of the Frontier

The previous pages discussed the dialectical evolution of the agribusiness frontier in the southern tracts of the Brazilian Amazon, from a national-developmentalist model in the 1970s to an agro-neoliberal model of agribusiness production since the 1990s, with a focus on the intricate configuration of new places and their evolution in time and space. The sensibilities of the poet Manoel de Barros informed our assessment of trends and contradictions of both the overall economic trends and local, place-based interactions. Considering the evidence presented, it can be concluded that this transition follows the introduction of new forms of public–private association, novel forms of socioecological exploitation and the suppression of alternatives to hegemonic agribusiness production. The national experience illustrates how agro-neoliberalism flourishes in a context of market-centred solutions and regulatory flexibility, but also that it demands novel forms of government support and relies on some of the oldest political traditions (e.g. aggressive manipulation of party politics, lack of transparency, deceitful claims of progress and elements of racism). The image of success is daily reaffirmed by sector representatives and endorsed by the national government in its effort to gain political support and maintain the export revenues generated by agribusiness. The various techno-economic innovations adopted by agribusiness players—including land and gene grabs, biotechnology and genetically modified organisms, dispossession of common land, financialisation and administration of production by TNCs—are all strategies that emerge from business and political interactions, which combine old and new features of the capitalist economy. The result is a nuanced and highly contested situation that connects, often in unexpected ways, different scales, sectors and public policies articulated around agro-neoliberalism.

Taking into account the complexity of achievements and failures, it is possible to verify that the true extent of agro-neoliberalism's success is highly questionable. The problems associated with the neoliberalisation of agriculture include a lack of access to affordable, nutritious food; the impacts of agrochemicals on communities and ecosystems; and the enormous concentration of power held by a small number of mega-supermarkets and agri-food corporations to control food production, distribution and consumption. While the neoliberal agribusiness sector has succeeded in crafting a positive image of technological and economic success, the federal government and the wider business community have become highly dependent on the export of primary commodities (to safeguard the national currency and avoid trade deficits, for example). Agro-neoliberalism evolves not only through attempts to influence the government, but also through further modification to the structure and rationale of the state. As a part of this turbulent and controversial process, new production areas are being incorporated with the employment of old and new practices of socio-environmental management and political legitimization. It is particularly in agricultural frontier areas, such as Mato Grosso, that the rationale of agro-neoliberalism is used to combine populist and neo-developmentalist traditions in order to disguise mounting impacts and inequalities. Ironically, when facing criticism from other social forces in the country, the agribusiness sector reacts with a pre-established rhetoric of heroism and entrepreneurialism that, in the end, serves the corporations and national politicians more than the farmers themselves. Agro-neoliberalism has been especially successful at the agriculture frontier because it is in itself an economic, ecological and ethical frontier, in which interpersonal and intersectoral relations have a particular configuration and impose undemocratic measures due to the primacy of production and the emphasis on rapid capital accumulation.

The first of the three future scenarios examined above—synthetically generated from the information acquired during the research in the country—suggests that agribusiness can continue to expand and further consolidate its influence as a strategic economic sector. This scenario echoes the current trend of growing soybean exports, conversion of pastures into monoculture grain fields and concentration of opportunities in the hands of large-scale farmers and associated transnational corporations. It basically represents the affirmation of a pre-cooked future that is politically used to justify the distortions and inequalities of the present. This mythical future, which is prearranged according to the opportunities and conveniences of agri-neoliberal agendas, has locked-in the narrow, anti-ecological production system. Furthermore, the rhetorical and material practices of agro-neoliberalism in Brazil represent the promise of a future that can never be fulfilled, as it brings back the worst features of the past (including legacies of slavery, over-exploitation of labour and resources, socioecological degradation and systemic violence used as a political tool). In the context of this scenario, it is difficult to envisage the emergence of any radical and genuine alternative that privileges social justice and the production of basic food. It seems that the two other plausible scenarios either comprise the return of more direct state interventions (to some extent, maintaining elements of agro-neoliberalising trends), or the collapse of agro-neoliberal agribusiness due to environmental and social tensions. Any other development would require a strong coordination of national and international social groups and political will to construct a different future for farmers, consumers and wider society. The future is wide open and will necessarily unfold according to complex social and socioecological interactions and to the evolution of the balance of power. Nonetheless, if in the end something like the first scenario prevails, it will be a disturbing demonstration that, because of the serious impacts of agro-neoliberalism in Brazil, the country will continue to harvest its exclusionary past and will be planting a narrow future for most.

Acknowledgements The content of this chapter is partially published by Springer under a CC BY 4.0 license in: Ioris, A.A.R. GeoJournal (2018) 83: 61. https://doi.org/10.1007/s10708-016-9754-7 © The Author(s) 2016. The chapter is also partially based on the author's previous publication: Antonio A.R. Ioris, Seeding a narrow future and harvesting an exclusionary past: The contradictions and future scenarios of agro-neoliberalism in Brazil, Futures, Volume 95, January 2018, Pages 76–85, ISSN 0016-3287, https://doi.org/10.1016/j.futures.2017.10.003 with the kind permission of Elsevier.

References

Agnew JA (2011) Space and Place. In: Agnew JA, Livingstone D (eds) The Sage handbook of geographical knowledge. Sage, London, pp 316–330

Alegria R (2005) The appreciative perspective of the future. J Future Stud 10(1):101–108

Antonsich M (2010) Meanings of place and aspects of the self: an interdisciplinary and empirical account. GeoJournal 75(1):119–132

Arenas HS, Geisse MG (2004) La Aproximación Humanística en Geografía. Revista de Geografía Norte Grande 31:31–52

Ash J, Simpson P (2016) Geography and post-phenomenology. Prog Hum Geogr 40:48–66

Asselin V (1982) Grilagem: Corrupção e Violência em Terras do Carajás. Petrópolis, Vozes/CPT

Barney K (2009) Laos and the making of a 'relational' resource frontier. Geogr J 175:146–159

Barros M (2013) Poesia Completa. São Paulo, LeYa

Barrozo JC (ed) (2010) Mato Grosso: A (Re)ocupação da Terra na Fronteira Amazônica (Século XX). Cuiabá, EdUFMT

Bartos AE (2013) Children sensing place. Emot Sp Soc 9(1):89–98

Bourgeois R (2016) Food insecurity: the future challenge. IDS Bull 47(4):71–84

Brando PM, Coe MT, DeFries R, Azevedo AA (2013) Ecology, economy and management of an agroindustrial frontier landscape in the Southern Amazon. Philos Trans R Soc B 368:20120152. https://doi.org/10.1098/rstb.2012.0152
Bridge G (1997) Mapping the terrain of time-space compression: power networks in everyday life. Environ Planning D 15:611–626
Buarque de Holanda S (1994) Caminhos e Fronteiras. São Paulo, Companhia das Letras
Cardoso FH, Müller G (1977) Amazônia: Expansão do Capitalismo. São Paulo, Brasiliense/CEBRAP
Chouinard V (1990) State formation and the politics of place: the case of community legal aid clinics. Polit Geogr Q 9:23–38
Churchill W (1997) A little matter of genocide: holocaust and denial in the Americas—1492 to the present. City Lights Books, San Francisco
Connor TK (2012) The frontier revisited: displacement, land and identity among farm labourers in the Sundays River valley. J Contemp Afr Stud 30:289–311
Cosgrove D, Jackson P (1987) New directions in cultural geography. Area 19(2):95–101
D'Incao MA, Silveira IM (eds) (2009) A Amazônia e a Crise da Modernização. Belém, Emílio Goeldi
Dardot P, Laval C (2010) La Nouvelle Raison du Monde: Essai sur la Société Néolibérale. La Découverte, Paris
Darling J (2009) Thinking beyond place: the responsibilities of a relational spatial politics. Geogr Compass 3:1938–1954
Das RJ (2012) Reconceptualizing capitalism: forms of subsumption of labor, class struggle, and uneven development. Rev Radic Polit Econ 44:178–200
Deininger KW, Byerlee D (2011) Rising global interest in Farmland. World Bank, Washington, DC
Dias EA, Bortoncello O (2003) Resgate Histórico do Município de Sorriso. Cuiabá
Dirlik A (1999) Place-based imagination. Review 22(2):151–187
Duram L, Oberholtzer L (2010) A geographic approach to place and natural resource use in local food systems. Renew Agric Food Syst 25:99–108
Dyck I (2005) Feminist geography, the 'everyday', and local-global relations: hidden spaces of place-making. Can Geogr 49:233–243
Elias D (2007) Agricultura e Produção de Espaços Urbanos não Metropolitanos: Notas Teórico-metodológicas. In: Sposito MEB (ed) Cidades Média: Espaços em Transição. Expressão Popular, São Paulo, pp 113–138
Folha de São Paulo (2009) Justiça de MT Decidirá Destino de Área que Vale mais de R$ 1 bi. http://www1.folha.uol.com.br/fsp/dinheiro/fi1209200918.htm
Foweraker J (1981) The struggle for land: a political economy of the pioneer frontier in Brazil from 1930 to the present day. Cambridge University Press, Cambridge
Galaz V, Crona B, Dauriach A, Jouffray J-B, Österblom H, Fichtner J (2018) Tax Havens and global environmental degradation. Nat Ecol Evol 2:1352–1357
Garcia R (2015) A Volta do Desmatamento na Amazônia. Scientific American Brasil 157:26–31
Harvey D (2006[1982]) The limits to capital. Verso, London and New York
Helfand SM, Levine ES (2004) Farm size and the determinants of productive efficiency in the Brazilian Center-West. Agric Econ 31(special issue 2–3):241–249
Ianni O (1981) A Luta pela Terra: História Social da Terra e da Luta pela Terra numa Área da Amazônia. Vozes, Petrópolis
Ioris AAR (2012) Applying the strategic-relational approach to urban political ecology: the water management Problems of the Baixada Fluminense, Rio de Janeiro, Brazil. Antipode 44(1):122–150
Ioris AAR (2013) The adaptive nature of the neoliberal state and the state-led neoliberalisation of nature: unpacking the political economy of water in Lima, Peru. New Polit Econ 18(6):912–938
Ioris AAR (2015) Cracking the nut of agribusiness and global food insecurity. In search of a critical agenda of research. Geoforum 63:1–4

Ioris AAR (2016) The politico-ecological economy of neoliberal agribusiness: displacement, financialisation and mystification. Area 48(1):84–91
Ioris AAR (2017) Agribusiness and the neoliberal food system in Brazil: frontiers and fissures of agro-neoliberalism. Routledge, London
Jepson W, Brannstrom C, Filippi A (2010) Access regimes and regional land change in the Brazilian Cerrado, 1972–2002. Ann Assoc Am Geogr 100(1):87–111
Lems A (2016) Placing displacement: place-making in a world of movement. Ethnos 81(2):315–337
Marx K (1976[1867]) Capital: a critique of political economy, vol 1. Penguin, London
Massey D (1993) Power geometry and a progressive sense of place. In: Bird J, Curtis B, Putnam T, Robertson G, Tickner L (eds) Mapping the futures. Routledge, London, pp 59–69
Massey D (1994) Space, place, gender. Polity Press, Cambridge
May J (1996) Globalization and politics of place. Trans Inst Br Geogr 21(1):194–215
Mcdowell L (2015) Roepke lecture in economic geography—the lives of others: body work, the production of difference, and labor geographies. Econ Geogr 91(1):1–23
McIntyre BD, Herren HR, Wakhungu J, Watson RT (eds) (2009) Agriculture at a crossroads: international assessment of agricultural knowledge, science and technology for development (IAASTD). Island Press, Washington, DC
Moran EF (1981) Developing the Amazon. Indiana University Press, Bloomington
Moreno G (2007) Terra e Poder em Mato Grosso: Política e Mecanismos de Burla 1892-1992. Cuiabá, EdUFMT/Entrelinhas
Öborn I, Magnusson U, Bengtsson J, Vrede K, Fahlbeck E, Jensen ES, Westin C, Jansson T, Hedenus F, Schulz HL, Stenström M, Jansson B, Rydhmer L, Öborn I, Bengtsson J, Jensen ES, Jansson B (2011) Five scenarios for 2050—conditions for agriculture and land use. Swedish University of Agricultural Sciences, Uppsala
Oliveira JM (1983) A Esperança vai na Frente: Contribuição ao Estudo da Pequena Produção em Mato Grosso, o Caso de Sinop. MSc dissertation, FFLCH/USP, São Paulo
Oliveira AU (1989) Amazônia: Monopólio, Expropriação e Conflitos. Campinas, Papirus
Oliveira AU (2005) BR-163 Cuiabá-Santarém: Geopolítica, Grilagem, Violência e Mundialização. In: Torres M (ed) Amazônia Revelada: Os Descaminhos ao Longo da BR-163. CNPq, Brasilia, pp 67–183
Pierce J, Martin DG, Murphy JT (2011) Relational place-making: the networked politics of place. Trans Inst Br Geogr 36:54–70
Pipino E (1982) Não há 'Inferno Verde'. Contato 33:3–6
Pocock DCD (ed) (1981) Humanistic geography and literature: essays on the experience of place. Croom Helm, London
Qian J, Qian L, Zhu H (2012) Subjectivity, modernity, and the politics of difference in a Periurban Village in China: towards a progressive sense of place? Environ Plan D 30:1064–1082
Raskin P, Banuri T, Gallopín G, Gutman P, Hammond A, Kates R, Swart R (2002) Great transformation: the promise and Lure of the times ahead. Boston MA, Global Scenario Group/Stockholm Environment Institute
Rempel ET (2014) Políticas Públicas Ambientais e seus Nexos com a Educação: Um Estudo no Município de Sinop-MT. Cuiabá, EdUFMT
Repórter Brasil (2010) Impacts of Soybean on the 2009/10 harvest. São Paulo, NGO Repórter Brasil
Santos JVT (1993) Matuchos: Exclusão e Luta. Vozes, Petrópolis
Sayago D, Tourrand J-F, Bursztyn M (eds) (2004) Amazônia: Cenas e Cenários. Brasília, UnB
Schmink M, Wood CH (1992) Contested frontiers in Amazonia. Columbia University Press, New York
Schutz A (1972) The phenomenology of the social world. Northwestern University Press, Evanston IL
Schwantes N (1989) Uma Cruz em Terranova. São Paulo, Scritta
Souza EA (2006) Sinop: História, Imagens e Relatos. Cuiabá, EdUFMT/FAPEMAT
Swyngedouw EA (1999) Marxism and historical-geographical materialism: a spectre is haunting geography. Scott Geogr J 115:91–102

Thornton PM (2010) From liberating production to unleashing consumption: mapping landscapes of power in Beijing. Polit Geogr 29:302–310

Thrift N (1983) Literature, the production of culture and the politics of place. Antipode 15:12–24

Tuan Y-F (1976) Humanistic geography. Ann Assoc Am Geogr 66:266–276

Vadjunec JM (2011) Extracting a livelihood: institutional and social dimensions of deforestation in the Chico Mendes extractive reserve, acre, Brazil. J Lat Am Geogr 10:151–174

Vieira PA Jr, Figueiredo EVC, Reis JC (2014) Alcance e Limites da Agricultura para o Desenvolvimento Regional: O Caso de Mato Grosso. In: Buainain AM, Alves E, Silveira JM, Navarro Z (eds) O Mundo Rural no Brasil do Século 21: A Formação de um Novo Padrão Agrário e Agrícola. Embrapa, Brasília, pp 1125–1155

Villas Bôas O, Villas Bôas C (1994) A Marcha para o Oeste: A Epopéia da Expedição Roncador-Xingu. São Paulo, Globo

Walker R, Simmons C, Aldrich S, Perz S, Arima E, Caldas M (2011) The Amazonian theater of cruelty. Ann Assoc Am Geogr 101:1156–1170

Chapter 4
Peasant Farming in the Amazon Frontiers

Et que sera un monde sans paysans?
Henri Mendras (1984: 363)

—Todo lo que se relaciona con el alimento del hombre es barbarie (…)
—Y más bárbaro lo que hacen: siembra de meíz para vender…
—Por eso el castigo…"
Miguel Asturias (1949: 231)

Abstract The chapter examines the main tendencies and perspectives of peasant family farming (PFF) in agricultural frontiers such as the Amazon. The ontological features of PFF are discussed, in particular, the multiple associations with, and subsumption to, agribusiness. It expands the discussion initiated in previous chapters to further discuss how, due to national politico-economic pressures, the Amazon was reinvented half a century ago as a vibrant agricultural frontier that attracted vast contingent of migrants due to coordinated government plans and, in more recent years, the cultivation of export-oriented crops. One very intriguing feature of this dynamic geography is that small family farming represents the 'other' of capitalist agriculture, but it functions as a hesitant form of alterity that both resists and fulfils rapidly expanding agribusiness. Furthermore, the intricate Amazon experience of approximation and distancing from agribusiness parallels with what has been happening with peasants all over the world. Small farmers maintain an 'inconvenient' and resilient autonomy that is not at all easy to erode; they exhibit an internal logic (beyond romanticism) and flexibility that is difficult for the reductionist operations of capitalism to incorporate. This interplay between survival, resistance and opportunities are even more intense in areas like the ones in the Amazon, where agrarian questions and the ability of peasants to respond to challenges are directly related to the range of inequalities and alliances that result from the specific circumstances of frontier making.

Keywords Food insecurity · Food sovereignty · Agroecology · Alterity · Subsumption

A. A. R. Ioris, *Frontier Making in the Amazon*, Key Challenges in Geography,
https://doi.org/10.1007/978-3-030-38524-8_4

4.1 Introduction

The intention here, following the preceding discussion of the advancing agribusiness frontiers in the Amazon, is to discuss the main tendencies and perspectives of peasant family farming (henceforth PFF) in the same socio-spatial context. There already exists a vast literature on the impacts of economic development and natural resource disputes between these two main spheres of agriculture, however, it is less common to find texts that synthesise trends and critically review the lessons learned. Most publications invoke and reinforce a basic agrarian model built around policy-making, internal migration, land management and mounting socio-economic inequalities. While those pressures remain important, there is insufficient attention to the politico-economic specificities of agrarian capitalism in situations of rapidly consolidating agricultural frontiers under the influence of hegemonic agribusiness (Ioris 2018a). Our departure point is the need to question the lived relations at the frontiers of agrarian capitalism, as in the case of Mato Grosso, where small family agriculture is not only under serious pressure towards specialisation, but also to creatively develop survival and reaction strategies. Instead of a focus on the rich and rapidly growing literature on agrarian trends, the goal is to understand the interface between different agricultural sectors in the context of regional development and landscape change. Agrarian capitalism cannot be narrowly blamed for the problems of PFF, but the main source of problems is actually the entelechy of the spatial frontier, which has evolved through a convoluted transition from pre-capitalist relations of production to dramatic processes of privatisation, commodification and financialisation. More recently, since the turn of the millennium, new spatial rationalities, combining an idiosyncratic agenda of environmental governance, poverty alleviation and neoliberalising practices, have deepened frontier making and further restricted the alternatives available for PFF.

There is an ongoing debate among social scientists about the need to properly consider the diversity of social groups in the Amazon, moving beyond the more common focus on indigenous groups to include various types of peasants (Nugent 2002). In that regard, our aim is to help to move PFF beyond the inadequate and derogatory consideration that it has often received. The material analysed in this particular chapter is based on a fieldwork campaign specifically dedicated to small family farming and focused on three agrarian reform areas: Gleba Mercedes, Zumbi dos Palmares and Terranova. The fieldwork comprised visits, interviews and other communication with farmers and family members, municipal and state authorities, union leaders, civil servants, rural school teachers, agronomists and researchers, cooperative administrators and politicians; also informal discussions and walks in the fields with family farmers and attendance of public meetings. Almost all those contacted were originally outsiders who migrated to Mato Grosso in search of economic opportunities and better livelihood conditions. The majority of present-day small family farmers moved to the Amazon or are first-generation descendants of poor migrants, increasingly intermarrying and interrelated with previous residents (most with indigenous ancestry).[1] The strategy was to learn about the specific geographical trajectories and historical agencies of peasants, avoiding normative ideas of capitalist development, but contrasting and comparing groups, times and spaces. First, it is necessary to revisit the ontological dialectics of approximation and distancing between PFF and capitalist agriculture.

4.2 Peasant Family Farming Dynamics

The role, the persistence and the prospects of peasant family farming—famously enunciated at the end of the nineteenth century as the 'agrarian question' of capitalism—have consistently intrigued many analysts. PFF eludes the reductionism of conventional political economy, typically focused on the primacy of the class struggle between proletarians and capitalists. At the same time, neoclassical economics underscores the supposed efficiency and aggregate value of agribusiness production. Yet, despite all the activity of modern, input-hungry agribusiness in large-scale farms, most food, jobs and valued relations in the countryside continue to be associated with PFF, which comprises 98% of all farms on the planet, occupies at least 53% of agricultural land and produces 53% of the world's food (Graeub et al. 2016). FAO and other multilateral organisations have accordingly highlighted its contribution to food security and regional and local development. That was further reaffirmed when the United Nations declared 2014 the International Year of Family Farming. However, PFF is not a straightforward concept due to the diversity of social formations and politico-economic conditions, including contrasting land sizes, economic outputs, technology, land ownership or insertion into markets (as pointed out by Shanin (2006), peasants make up the majority of humankind and this will remain the case well into the current century). The very expression 'family farming' is a misnomer, considering that PFF is neither necessarily conducted by a family, nor only about land cultivation (it increasingly involves a number of other economic and non-economic activities). PFF is actually becoming a post-family activity, even in very low-income areas like the Amazon, as it is common for wives and children to live in cities in order to have access to education and urban jobs. Given that PFF increasingly involves other sources of income, although food production remains important, it also often involves more-than-agricultural activities.[2]

---

[1]PFF in Brazil today is a diverse category that includes European descendants, slave descendants (*quilombolas*), squatters or partners in large farms, agrarian reform settlers, indigenous groups and traditional extractivist communities, who together represent 84.4% of the rural households in Brazil (4.37 million units) but who hold only one quarter of agricultural land (80.25 million ha), according to Medina et al. (2015). 62.31% of the family farmers in Brazil have a total income of less than the minimum wage and no more than 12.77% benefit from agricultural policies; 80% of family farms consist of less than one fiscal module (the average property is 18.37 ha).

[2]The Brazilian Law of Family Farming (Law 11,326/2006) defines family farms as small units (less than four fiscal modules, defined locally), reliant mainly on family workforce, with income

PFF consists of a broad conceptual envelope that contains a diversity of groups, from relatively wealthy producers in Western countries, to indigenous communities in the Andes, pastoral tribes in Africa and rural societies in Asia.[3] According to Wolf (1966), what differentiates peasants from 'primitive societies' is the production of a surplus, beyond bare subsistence, which is normally appropriated by the dominant political group. However, it is not easy to come up with a satisfactory conceptualisation that embraces the peasant economy, family farming and peasant labourers. Many definitions of peasant farming seem largely tautological, as they tend to presuppose most of what needs explaining, such as the daily operation of family farming and the important insertion in property relations and market transactions that define an increasingly capitalistic world. Although the ownership of land (differentiating peasants from proletarians, who do not own the means of production) and the centrality of family labour (typically unpaid, although there is occasional use of paid labour) are commonly used to define the peasant economy, there are several obstacles to understanding small family farming in the contemporary, capitalist world. One is that family farming, in the form of family units of production, predates capitalism (the normal analytical model has to do with European feudal relations, but comparable systems of production existed and continue to exist around the world). A related difficulty is that both capitalism and agriculture in the twenty-first century are, to a large extent, different than in the early and mid-twentieth century, when most of the authoritative literature on the agrarian question was published. Capitalism remains capitalism, but the strategic connections between state and private operators have evolved significantly. This is particularly true in relation to state institutions that guarantee and legitimise land and nature grabbing, but also in relation to various forms of national and international assistance available to family farmers.

As is well known, the literature devoted to the agrarian question and small family farming resurfaced in the 1960s and triggered a massive debate about class identity, ways of life, economic specificity, units of production, land tenure and cultural formations (an overview can be found in Brookfield and Parsons 2007; Shanin 1987; Wolf 1966). Critical scholars have oscillated between the so-called Marxist and Chayanovian perspectives, with multiple variations, while Bernstein et al. (2018) emphasise the need to incorporate new topics, such as gender, culture and ecology. The majority of the publications, revised by Garner and O Campos (2014), focus on size of property, kinship relations, diversification, family land ownership, labour use and management, production basically limited to meeting subsistence needs, the indivisibility of production among family members, and pluriactivity, as well as the supposed environmental, social and cultural functions of small family farming. It is common to try to explain family farming from the perspective of farming in

predominantly from the farm and management by the peasant family. This legislation has important similarities with the legal definition of PFF in countries such as Argentina, Chile, USA and Uruguay.

[3] Interestingly, La Vía Campesina, which is the most emblematic mobilisation of peasants in the world today, never attempted to resolve these conceptual controversies; rather, the social category 'peasant' is taken for granted and treated politically as a unifying concept.

general, making use of criteria such as productivity, levels of investment, mechanisation, economies of scale, land property and opportunity costs (e.g. Van Vliet et al. 2015). Some in the North American tradition even differentiate between subsistence or peasant farming (considered backward and decadent) and family farming (which is characterised as more receptive to 'modern' technologies and intensification of production), while there is often also particular attention to sources of income, which are expected to be predominantly based in the farm unit, and reference to PFF's joint ownership of the means of production and labour power.

It is important to identify the core elements of this controversy about the pillars of peasant family farming. According to Friedmann (1980), PFF can be situated as a process, in a perpetual state of flux, but necessarily retaining a configuration, rationality and positionality that contrasts with capitalist agriculture. The key explanation, thus, lies in the 'form of production' simultaneously based on the internal characteristics of the farming unit and on the external characteristic of the social formation. Barta (2011) agrees that today there are not two, but only one mode of production—capitalism—which dominates and integrates the whole economy (this directly contrasts with Chayanov's description of a unique peasant economy (cf. Chayanov 1966, 1991)). PFF is characterised by non-capitalist internal relations, but is fundamentally subject to capitalist rules (especially the valorisation of capital) and, more importantly, it results from the very reproduction of capitalism. Brookfield and Parsons (2007) make clear that family farmers retain individual control over the land they work (as owners or tenants), produce commodities both for sale and for subsistence, utilise mainly family labour mobilised through kinship and interdependent relations, and have a partial engagement with markets. Likewise, for Shanin (2006), small-scale farmers produce mostly for their own consumption and for the fulfilment of political and economic obligations. This group is, therefore, characterised by four independent, but mutually reinforcing facets: (1) the peasant family farm is a multidimensional unit of social organisation; (2) land husbandry is the main means of livelihood; (3) there are specific cultural patterns linked to the way of life of a small rural community; and (4) the 'underdog' position, that is, the domination of PFF by outsiders.

This fundamental dialectic between capture by capitalist relations of production and an internal rationality that is more-than-capitalist (and, quite often, anti-capitalist) has great analytical relevance to understand the insertion of PFF into an agri-food sector increasingly dominated by the agribusiness model. PFF is affected, harassed and undermined by agribusiness, but without being completely absorbed by market-based relations (to the extent that family farmers are able to resist the pressures). The contested landscape of contemporary agriculture and the growing reliance on agribusiness-based production systems has resulted in high levels of unsustainability and high health and environmental impacts (De Schutter 2017), which leads to uncertain prospects in terms of public health, rural development and food security. Although there are many intersections and hybridities between these two main agri-food paradigms, a significant proportion of urban and rural populations around the world increasingly consume low-quality food produced with the intense use of agrochemicals, genetically modified crops and heavy machinery (Clapp 2018). Mainstream agri-food systems and governance mechanisms are highly path-dependent, and resistant to reform, which indicates the need for critical and innovative studies to question the basis and consequences of agriculture dualisms unfolding at different scales and with complex repercussions for consumers, communities, ecosystems and national economies. In broad Gramscian terms, peasant family farming provides the basic elements of resistance and contestation that inform a critique of the powerful ideological and material praxis of agribusiness hegemony. Furthermore, PFF plays a very important role in the pursuit of food security and food sovereignty, which have been significantly undermined by the advance of intensive, input-hungry agribusiness. In a context where food, and the way we experience it is increasingly disfigured and metamorphosed into something new, the question of food security and the role of family farming go beyond issues of access, affordability and self-sufficiency. The semiotics and the social construction of relations around food suggest that the agenda of food security needs to incorporate and react to many layers of complexity, in particular, family farming, as discussed next.

4.3 Food Insecurity and Peasant Family Farming

The contemporary debate on food insecurity and the role of peasant farming is certainly not trivial. There are regular, almost daily news stories about food, hunger, trade, safety, and so on somewhere on the planet, which seems to annul our ability to think critically and react to mounting demands and distortions. Approaching this from a critical perspective, it can be argued that food insecurity is, basically, the result of the concurrence of two main driving forces unleashed by growing consumerism and commodification: first, the reduction of agriculture to agribusiness and, second, our disconnection from the sources, provision and content of our food. In relation to the first main driver of food insecurity, for many thousands of years agriculture was not merely a human process employed to produce food and raw materials, but the expression of particular cultures, the manifestation of accumulated knowledge and the basis of social relations taking place in specific socioecological settings. With the advance of capitalist relations of production, profound changes took place, initially in the countryside of Western European countries, and then most of the world was inserted into the sphere of influence of modern capitalist agriculture. The process involved, in general terms, the appropriation and privatisation of land and other commons, the expulsion of large contingents of the population to work in urban areas, and the specialisation of production according to the demands of industries and cities. In the history of capitalism, agriculture and farmers have always taken a twofold subordinate position, combining the subordination (exploitation) of labour to capital and the countryside to the city. This phenomenon can be described as 'a movement from food-as-nutrition and agriculture-as-social-integration to a situation in which agri-food operations are carried out primarily to circulate and accumulate capital.

The ideological and practical reduction of food to the realm of commodities, exploitation and profit also represents a decisive barrier to the resolution of nutritional and environmental problems. In synthetic terms, this is a gradual shift, which began at the end of the nineteenth century, from *agriculture-cum-food* to *agriculture-cum-agribusiness* (Ioris 2017: 21). Due to scientific developments in modern agriculture—which, incidentally, are meant to transform the ecological conditions of the soil and dramatically alter the genetic makeup of organisms, instead of focusing on the much more sustainable adaptation of plants and animals to their ecosystems—in the last few decades agricultural technology has exponentially accelerated the erosion of social relations and food as life enrichment in favour of agriculture as business and profit. It is very disconcerting that what is normally described as modern and superior agriculture is in effect a narrow and impoverished form of agriculture, determined by market rules and political bargains that, in the end, cause and perpetuate food insecurity. As demonstrated by Clapp and Fuchs (2009), food systems around the world are increasingly affected by corporate-dominated global systems, and corporations are involved at all stages along the production chain, compromising small farmer livelihoods, environmental quality, food security and consumer sovereignty. The result is that *agriculture-cum-agribusiness* is fundamentally anti-food, anti-people and anti-future, as it is inescapably led by the logic of the circulation of capital and short-term profits. Food insecurity derives here from the mistaken delegation of something so basic and precious as food to the exclusive realm of moneymaking. One madness of the world today, among many others, it to transfer the control and the choice of what we eat or ingest into our bodies to nameless players with explicit but selfish agendas: plantation farmers, supermarket managers, scientists, bureaucrats, and unelected regulators, among others.

Food insecurity is a clear sign of the highly risky and limited perspectives of the increasing capitalisation of agriculture; in simple terms, the more agribusiness, the higher the level of food insecurity. The influence of huge oligopolistic food producers and distributors is facilitated by the intervention of neoliberalised governments under the ideology of market-based efficiencies (Otero et al. 2018). Food insecurity as such (unlike the similarly tragic periods of food scarcity experienced in pre-capitalist times) is something that can be found only in the capitalist world. More than that, food insecurity is the result of capitalist relations of production and reproduction based on the perpetuation of symbolic and material forms of scarcity, which serve to maintain the primacy of exchange-value over other values and to justify exploitation as the basis of profit-making. Just as liberal capitalism has increasingly undermined liberal democracy (because of the internal contradictions of capitalism, the drive for expropriation, exploitation, commodification and accumulation), an agri-food system dominated by financialisation trends does not permit genuine prospects for fair food security. It is ever more noticeable in the early twenty-first century that liberal democracy, as adopted in Western countries and replicated around the world, suffers from serious problems of representation, lobbying, manipulation of parliamentarian procedures, sidelining minorities through political correctness, electoral fatigue and indifference. The relation between people and elected politicians is worryingly mediated by the influence, the values and the language of large business and transnational banks and corporations (the British case is emblematic, due to the enormous dominance of the banking sector in the national and, particularly, London-based economy). In 2016 the USA electoral process produced one of the most troubling aberrations with the election of a controversial billionaire for president (himself a famous consumer of fast-food who has been involved in sex scandals and suspicious businesses practices), representing the direct and unqualified takeover of democracy by large business. A subsection of that trend is the reduction of agriculture to agribusiness, food to processed, commodified food, and eaters to supermarket consumers. Following this logic, it is not surprising that the hegemonic thinking about food security is pervaded by quick-fix solutions that cancel the possibility of rupture and, ultimately, perpetuate the causes of insecurity.

The second main driver of food insecurity is a direct extension of the first: food has become something abstract, taken for granted and disconnected (disembedded) from people's basic concerns, given that the problem is basically reduced to a purchase operation. As long as people have money, they assume that some sort of food will be available to buy. This simplistic, monodimensional thinking about food is expanding not only in urban areas but also affects farmers and rural dwellers who increasingly rely on local vendors (supplied by corporate-controlled supply chains). The aberrant consequence is that the question of food is equated to the possession of money, given that very few people (farmers in Western countries included) cultivate what their families eat. The majority do not have the knowledge or the means to cultivate something for their own kitchens. Global society is increasingly separated and unaware of the origin, substance and practices behind what we eat. People's experience is, in most cases, limited to supermarket shelves, local markets or fast-food restaurants. This condition obviously leaves low-income people very vulnerable to the vagaries of the market, as in the case of the food crisis that followed the crash of global markets in 2008 when food prices doubled or tripled before eventually returning to the pre-crisis level of 2006 (Magdoff and Tokar 2010). In sociological terms, this problem is directly related to both the concentration of power in the hands of corporations and supermarkets and the centuries-long process of peasant displacement and land commodification.

From the above, it can be seen that the problem of food has been increasingly dislocated from the mere sphere of farms and local markets and, consequently, food insecurity has much wider repercussions. Different perspectives compete for prominence and policy-making influence. The most common reaction to the pressures of food security is the productivist paradigm. Basically, the argument goes, if there is a lack of food, the answer is to produce more and make food more easily available in time and space (while issues of distribution and socio-spatial inequalities are much less important). The narrative is that because of steady population growth and less significant productivity gains, there is a growing threat to the ability of entire countries and regions to feed themselves, but with rising prices of basic food commodities, the response has been to produce more. An important facet of the productivist response to food insecurity is the claim that attention should be focused on technology, machinery, making more land available for cultivation and increasing inputs to produce higher outputs (especially in terms of caloric output). In that way, agriculture's cultural and ecological dimensions are disregarded. Agriculture is, in practice, equated to biochemistry, engineering and business administration, as much as food is reduced to a long table of nutrients rather than an important element of family and community relations. Those excuses pave the way for technological fixes that, in the end, represent a form of Malthusianism by the back door. Carolan (2013: 2) rightly surmises that food security 'has been hijacked by a vocal and powerful minority' who translate it into narrow, productivist policies.

With the declining economic role of agriculture in so-called post-industrial countries, a new argument started to emerge, with scholars starting to suggest that agriculture was embarking on a post-productivist or multifunctionality phase, much as the environmental question had reached an ecological modernisation stage and the economy a service-based status. According to this (mainly British) literature, while productivist agriculture is intensive, expansionist and industry-driven, constantly trying to boost production of single crops over large areas, dependent on state support and aiming to reach national self-sufficiency, a post-productivist model (triggered by regulatory reforms, such as the adjustments of CAP-related policies) focused on diversification, pluriactivity, quality rather than quantity, and environmentally friendly agriculture. It also includes encouragement of activities that treat the countryside as something to be 'consumed', through activities like recreation, tourism and the conservation of natural and cultural heritages (Ilbery 1998). The arguments in favour of a post-productivist, multifunctional agriculture has parallels with the search for more sustainable farming. An alternative to the high-impact productivist approach, post-productivism has been described as the pursuit of sustainable farming systems (Wilson and Burton 2015). Solutions to ensure food security should then observe sustainability concerns and be more people-centered, avoiding the degradation of systems and their ability to meet human needs in the long term (McDonald 2010). Sustainability authors denounce the real costs of 'cheap food' as the super-exploitation of workers and the environmental passivity produced by mainstream agriculture but not incorporated in the selling price. The sustainability of agriculture is related to the capacity of an agroecosystem to maintain production through time, that is, remain stable under a given set of economic and environmental circumstances (Arulbalachandran et al. 2013).

This second position is consistent with the growing realisation that the agri-food systems developed over the past half-century have proved to be clearly unsustainable, resistant to reform and fraught with problems, such as their failure to reduce rural poverty and address power imbalances in food chains (De Schutter 2017). At face value, it seems that the shortcomings of input-intense agriculture are self-evident (at least to a large proportion of those questioning its long-term prospects) and that sustainability could be achieved with a more diversified type of agriculture that would deliver high-quality goods and services with lower use of chemical inputs. In practice, however, calls for sustainability are often undermined by the multiple continuities between 'conventional' and 'alternative' production systems, leading to uncertain transitions from one model to the other (Walford 2003). As has happened in many other areas of social activity, there was some intellectual space for self-criticism, but this inevitably replicated the same dualist and utilitarian biases that informed the productivist mindset. Basically, sustainable food entails a more efficient use of resources and the mitigation of negative impacts on natural and human capital. It is about striking a balance between the production of goods and services and the rate of natural resource consumption, but it is far from obvious how this might be achieved, and the debate has become depoliticised (Redclift 1997). Sustainable food is often associated with organic markets, where local farmers can meet their customers on regular basis, but calls for food and agriculture sustainability in isolation from more fundamental changes, just like post-productivism, are fraught with contradictions and shortcomings.

A third main response focuses on the ethical and human rights dimensions of food security. The is epitomised by calls for food justice, which entail more than mere distributive action but also require the recognition of political voices and the widest possible representation in decision-making and public policies on production, consumption and distribution of food, as well as the normative connections between food-related practices and collective self-determination (Scoville 2015). It is not difficult to perceive the cross-references between food justice and environmental justice, which both demand the fair treatment of all people, regardless of race, colour, national origin or income, with respect to the development, implementation, and enforcement of laws, regulations, and policies. The departure point of this third perspective is the abundant evidence around the world of the connections between poverty, social inequalities and food insecurity. Food insecurity is thus interpreted as a sociopolitical problem in particular historical and geographical contexts, which cannot be resolved with better productivity and market globalisation. Because of the accumulation of state and market failures, hunger and malnutrition are not distributed randomly but concentrated in certain areas and among certain groups. For instance, UNICEF (2017) estimates that nearly half of all deaths in children under five are attributable to undernutrition, translating into the loss of about three million young lives a year (associated with the increased frequency and severity of infectious diseases); despite important gains in recent decades, high levels of mortality continue to be found particularly in South Asia (48 per 1,000 live births) and Sub-Saharan Africa (78, with a peak of 95 in West and Central Africa).

Food justice has paved the way for farmers' markets, as well as agroecologically managed farms, community-supported agriculture initiatives, fair trade and organic food suppliers, etc. which are all attempts to connect producers and consumers and highlight the importance of place-based agri-food chains. However, there is a common risk of local farmers' markets becoming restricted to better-informed, high-income shoppers in search of better-quality food, but oblivious to wider food insecurity trends and the nutritional challenges faced by the majority of the population. The romanticism of localised food chains leads to serious biases and represents a form of elitism by the back door. A similar distortion, which also tempered the impact of the food justice agenda, was the appropriation of food sovereignty by the International Assessment of Agricultural Science and Technology for Development, an intergovernmental panel under the sponsorship of the United Nations and initiated by the World Bank in 2002. Its synthesis report describes food sovereignty as 'the right of peoples and sovereign states to democratically determine their own agricultural and food policies' (IAASTD 2009), which is a rather technocratic definition, empty of the political content that characterises the activity of La Vía Campesina and other critical grassroots movements.

This indicates the role of a non-essentialist interpretation of multiple scarcities and the need to address the totality of the experience of scarcity (Ioris 2012). An improved food security condition depends, first of all, on the ability of peasant family farmers (the real producers of food, cf. Shiva 2016) to maintain their activity in the long-term, which is related to problems such as land access, fair market transactions, just payment of rural labourers, the removal of control over food by supermarkets and transnational corporations, the enhancement of peasant associations, etc. The reaffirmation of agriculture as primarily a source of food, not profit, is evidently a major and difficult political question. In addition, food security depends on the offer of a 'full plate', not only in terms of quantity, but quality (in terms of nutrition, safety, freshness and in accordance with local culture and the autonomous wishes of producers and consumers). A full plate is, in fact, a perfect metaphor for the totality of relations underpinning food security. Good food is 'just food', rather than products manipulated and commercialised by corporations. Food security requires rethinking food and revisiting agriculture, which ultimately requires rethinking the world. These two mega-agendas will have to incorporate cultural, economic, agrarian and technological differences, but in general terms the task is similar in countries of both the Global North and the Global South. However, the non-Western parts of the world seem to have much more to offer and a lot to teach. This is perhaps the moment to return to common sense and appreciate food as food.

Having briefly examined the tensions and interconnections between family farming, food insecurity and agribusiness farming, our focus in the following sections returns to the distinctive features of Amazonian agricultural frontiers, as in the case of Mato Grosso, where both agricultural categories have been growing, despite many politico-economic obstacles, as crucial components of the wider process of frontier making promoted by the Brazilian state since the 1960s. In terms of the global division of agricultural labour, Brazil is expected to remain a producer of cheap commodities; this is primarily guaranteed by the availability of land, water and resources at the socio-spatial frontier in the Amazon. This geopolitical arrangement certainly fails to resolve the overall problems of food insecurity and malnutrition, but it is also a form of 'silent violence' (cf. Watts 2013) whereby the world capitalist economy flourishes due to a combination of exploitative agriculture, land grabbing and resource and land rents. As mentioned above, the frontier is a new spatial setting, but its novelty is partial and qualified, considering that it brings back the worst of the past lived in the original economic centres. The interplay between 'core' and 'frontier' areas is revealed through the perverse 'law of scarcity-abundance', which synthesises the general tendency to deal with mounting scarcity in central areas through the pledge of abundance at the frontier, although in practice new rounds of scarcity emerge in both areas due to the spatial dynamics of capitalism (Ioris 2018b).

One of the key players in the geographical dynamics of the agricultural frontiers in the Amazon (in addition to large landowners, land speculators, military engineers, etc.) was the small family farmer—known in Brazil as the colono—who partly replaced, and inherited important technological and sociological legacies from, previous forms of peasantry established during colonisation or at the time of the 'rubber boom' (see Adams et al. 2009). Colonos are individuals who attempted to take part in the dream of a better life in the frontier, but who in the majority of cases failed to see that dream realised. As much as agribusiness, frontier peasants are the product of regional development promoted along the lines of the configuration of new frontiers. The mechanisms of convergence and differentiation will be analysed in the next section.

4.4 Peasant Family Farming in the Amazon's Socio-Spatial Frontiers

The Amazon continues to capture the global imagination and to stir debates about regional development, natural resources, socio-cultural impacts and environmental degradation. It is today one of the crucial arenas for the expansion of global capitalism and experimentation with national economic agendas. One of its most striking geographical features is the perennial opening and closing of spatial frontiers associated with social, political and ecological struggles. Particularly in the Brazilian section, land has always been an object of dispute in the Amazon, mainly due to the availability of 'natural' riches and the expectation of future rents from land speculation (not so much from production). To a large extent, this has followed the wider controversies about land access and social inclusion of rural populations in Brazil, but there are also specific attributes to the agrarian question in the Amazon—where PFF was relocated and transformed as part of frontier making. For a long time, there was scant recognition of the existence of the peasantry in Brazil due to the prevalence of monocultures and extractivism since the colonial period (Caldeira 2017). Gradually, more authors have started to pay attention to the economic relevance of small production units—the free peasantry—functioning side by side with plantation farms, as well as cattle ranches, and their role in the supply of food for the latifundia and the urban settlements (e.g. Queiroz 1976).

Academics have also been motivated by the intricate campaign for agrarian reform that occupied a central position in the Brazilian political debate during most of the last century. The struggle for land was part of the wider confrontation of class privileges by those historically excluded from the country's modernisation. As in other countries, PFF did not easily fit into (conservative) plans for industrial growth and social transformation launched in the first decades of the twentieth century. After great political turbulence in the 1950s and 1960s, fuelled by the operation of the Brazilian Communist Party (PCB), a military regime took power on 1st April 1964 and retained the direction of agrarian policies in favour of the old elites. The regime made a small gesture towards public opinion with the introduction of the Land Statute, but in practice the priority was the support of commercial, large-scale farming and the promotion of colonisation projects. The dictatorship pushed forward an agenda of national security, infrastructure expansion and land grabbing in the Amazon, funded by foreign loans and rationalised by Cold War fears of foreign invasion and socialist upheavals. It was basically a crude process of social engineering that attracted peasants and workers to colonisation projects along the new motorways, while at the same time the state was subsidising large-scale land grabbing by corporations and private individuals.

The initial formation of the Amazon peasantry (with the characteristics recognised by Shanin 1987) actually occurred due to the arrival of settlers from the Northeast Region during the so-called rubber boom (1860–1920), which led to miscegenation with the indigenous population and shaped heterogeneous social groups known today as *caboclos* or *ribeirinhos* [riparian dwellers] (Hébette 2004). The intermingling of different groups produced what Nugent (1993) describes as 'hyphenated' (hybrid) identities. However, the military regime post-1964 introduced regional development plans that led to the gradual incorporation of the Amazon (considered in the official discourse as 'no man's land') in a conservative and elitist process of economic modernisation (that culminated in recent years in the consolidation of agribusiness and the construction of large hydropower schemes). In the 1970s, attempts to develop the Amazon brought to the region a new character, who had barely existed before: the colono, a new type of peasant primarily associated with public or private colonisation schemes. Most colonos had Italian and German ancestry and moved from the southern Brazilian states, especially from Rio Grande do Sul, due to family poverty and land scarcity. According to Foweraker (1981: 68), the colono 'moves when the farm can no longer provide subsistence for all members of the family', even if they still hold some small tracts of land. The expulsion of the rural poor was also related to the advance of the Green Revolution and the growing scarcity of land for the reproduction of PFF in Brazil's southernmost state of Rio Grande do Sul.

Despite the rhetoric of opportunities and the promise of a better future, the frontier was characterised by wastage of resources, socioecological impacts and high levels of violence and criminality, which suggests that the main purpose of the frontier was not agricultural or economic, but political and ideological. It was formed—at a high cost in terms of public money, local ecosystems and the hardship of most people involved—fundamentally for the extraction of economic and political rents (Ioris 2016). It was part of the construction of an image of Brazil as a major economic and geopolitical player, and the government used, and continues to use, regional development to increase its political legitimacy and mitigate politico-economic demands in the consolidated areas in the south and east of the country. The frontier was an important component of the wider mobilisation of labour and resources to increase a national version of global capitalism that was basically a compromise between old and new economic elite groups. It was an orchestrated attempt to divert political pressures away from disputed areas and to minimise the growing contradictions of conservative economic policies. As discussed by Watts (2013), in a milieu of not fully developed capitalism, the economy in the region was not separate from the state, but the two overlapped. There was no relevant economic activity outside the sphere of direct state influence and all processes of production, commercialisation and consumption depended on the action of the state (this started to change in later decades with the deeper commodification of nature and market-based values and relations, see below). In the receiving areas, such as Mato Grosso, a similar socio-spatial hierarchy was set in place (which largely replicated the configuration in the original areas) with large-scale landowners at the top (cattle ranchers or planation farmers, typically receiving generous state subsidies, free land and very friendly contractual terms) and a vast mass of peasants (colonos) installed in agrarian reform schemes (private and public) along the roads being opened through the forest.

The experience of Terranova, in the north of Mato Grosso, illustrates the many ramifications of the attempt to organise new peasant groups in the context of the agricultural frontier. The Terranova Programme, described by Santos (2008), was organised in 1978 by the cooperative Coopercana and aimed to relocate peasants from Rio Grande do Sul after they were forced out of the Kaigang indigenous reserve (illegally occupied 20 years earlier). It benefited from the construction of the BR-163 road (Cuiabá-Santarém), a large engineering project that had devastating consequences for the local indigenous Kren-Aka-Rôre people. It comprised 435,000 ha divided into 1,600 lots, initially with 100 ha each, later reduced to 50 (only half the land could be deforested), plus 2 ha for residents in six agro-villages. The new farmers received support and credit from the government, but there was growing frustration in relation to the unfulfilled promises, logistical difficulties, lack of public services and absence of market. There were also disputes between farmers and the cooperative, as well as competition between families. Soon the farmers were heavily in debt and struggling to survive. 80% of those who arrived in the first phase of colonisation in 1979–1981 returned to Rio Grande do Sul. Those who managed to remain in the area secured the dream of their own piece of land, but the reality was different than the original plan and the cost much higher. Desconsi (2011), who conducted ethnographic work on the families of southern migrants who moved to Mato Grosso, also demonstrates that those families who remained on their land improved their material circumstances, but without changing their subordinate social condition. Although marginalised in sociopolitical and economic terms, these groups of migrant colonos continue to be recruited to bring progress to Mato Grosso, and invoked to legitimise the agricultural frontier.

What happened in Terranova was similar to most other colonisation schemes in the Amazon, where there was too much reliance on a capitalist space produced by the state. The footprint of land use was primarily based on the expansion of the area, taking place in cycles related to waves of government support, rather than on the intensification of production (Brondizio 2009). The frontier was conceptualised and promoted as a realm of opportunities for all, while in effect the main driving-force of the frontier was the relatively easier access to public funds. Both peasants and large-scale farmers, despite major class differences, shared this same ambition: move to the frontier in order to benefit from the availability of concessions, land and resources. In the process, not only was the state apparatus pervaded by the subjacent agrarian question, but also, more importantly, the process of agrarian change reflected disputes and power imbalances between large-scale and peasant farmers that directly shaped state action. Legislation, official plans and public policies had to respond in different directions and ended up bringing the old agrarian disputes from elsewhere to the Amazon. The organisation of PFF at the economic frontier was basically the result of external processes shaping the region, such as internal migration, national development policies and the transference of social institutions from core politico-economic areas. After the colonos settled in the region, there was a gradual and partial commodification of products, labour and land in highly heterogeneous ways. In periods of production crises, there was a resurgence of old practices and ancient techniques brought from the migrants' original areas, mixed with regional practices that they had to learn from those already living in the area (such as the use of resources from the rainforest, the cultivation of local plant species and adapted housing construction techniques).

It is quite remarkable that in the process of frontier making formal land ownership was not necessarily the only or even the main goal. In a context of uncertain rules and fuzzy property boundaries, land tenure tended to be tentative and conditional. The same government that encouraged people to come to the frontier was unable to guarantee steady assistance or fully enforce private property institutions. Campbell (2015) describes the genesis of private property in the Amazon as something that is 'conjured' and that both reconstructs the past and anticipates the future. Nothings is really settled, but land tenure can evolve according to political disputes and the emergence of opportunities. Government action helped to create the contingency of land ownership, considering that the main objective of the military technocracy was to reduce social tensions in the south (as in the case of those who migrated to Terranova) and foster a new developmentalist cycle through tax reduction associated with projects in the Amazon. Those holding official property titles can be forced out when someone stronger appears with a similar document granting ownership of the same area. Small-scale farmers had to keep their options open, frequently moving to other locations or enlisting in new government projects.[4] Figure 4.1 shows an area recently occupied by a PFF in an agrarian reform project near Sorriso. Their main aspiration was not merely land ownership (something historically difficult to secure and maintain for peasants in Brazil), but the realisation of the frontier dream, whatever that meant and according to the circumstances of each family, time and location. That produced high-level of risk and the need to be part of and operate through networks to develop a range of survival strategies, including jobs outside the property and time spent as wildcat gold miners (from time to time a new mining area is located and attracts large contingents of people).

The sociopolitical landscape of PFF at the frontier was, and continues to be, a story of struggles, advances, alliances and setbacks that evolves according to special agrarian capitalist circumstances (for the uncomfortable sense of misplacement underpinning the agricultural frontier, see Ioris 2017). The ability of the state to fund public schemes quickly deteriorated and was diminished by the macroeconomic

[4]In more recent years, family farmers have even incorporated new discourses, such as environmental conservation, in order to make the most of government projects and internationally-funded schemes.

Fig. 4.1 Agrarian reform plot being converted into agricultural production

turbulence during the 1980s. Colonisation projects introduced in Mato Grosso struggled to survive and entered a phase of economic and organisational decadence. The colonos coped with the odds against success in smallholder production and tried to respond to official programmes and incentives as best they could, but such initiatives were often discontinued or mismanaged, undermining the colonos' ability to act 'rationally' (Brondizio 2009). Many lost their land (regularly or unofficially occupied) and either returned to their areas of origin, moved to other locations or were proletarianised. Those who maintained a rural property incorporated capitalist features at different levels, establishing particular forms of engagement with the regional economy and the political system. Also, in some areas around the urban centres, better-equipped groups of family farmers specialised in certain crops (such as vegetables, irrigated maize, milk, etc.) and managed to gradually accumulate capital and purchase nearby land. Specialisation of peasant farmers reveals the pressure to 'modernise' along the lines of input-intense technologies, specialised production and growing market insertion. The modernisation of PFF was particularly nurtured by the National Programme of Family Farm Support (PRONAF), established in 1995–1996, as an attempt to advance a new paradigm of rural development through easier access to credit and various other support schemes.[5]

[5]In practice, the history of the programme has revealed uneven results, unremitting operational problems and aggravated regional and intrasectoral inequalities (Wanderley 2014).

Public policies and legislation introduced in the last two decades, such as PRONAF, have essentially tried to bring PFF into the domain of mainstream agriculture through the emergence of the new category *agronegocinho* [literally, 'small agribusiness'] in an attempt to capture the concerted ideological attempt to redefine PFF. Even the president of the union of rural workers in Sinop affirmed 'our future is to make the most of the agribusiness opportunities available for family farmers.' In another interview with a senior researcher from Embrapa (the federal public agricultural research corporation) in Sinop it was argued that the main problem of family agriculture in Sinop is the 'lack of entrepreneurial inclination' [*falta de aptidão empresarial*]. There are apparently great opportunities to improve the production of fruit and vegetables in Mato Grosso, considering that more than 80% is still imported to the region, but the argument is that most local peasants are unprepared to run their farms as competitive businesses. For the researcher, the key problems faced by PFF are not related to the ability to produce or the conditions to produce (land, water, technology, labour), but to the need to stimulate production champions, leaders, as models for others to follow and to reduce dependency on the state. According to the same interviewee, 'the frontier situation forces small family farmers to become entrepreneurs.[6] In a graphic interview with the head of rural development for the Sinop city council, it was added that

> Peasants continue to struggle, what is lacking is more professionalism, investment in food supply chains, less politics and more economic autonomy in the producers, who soon abandon [farming] and move to the cities… Agrarian reform projects are archaic, only focused on land allocation and not on production; we cannot give land for free, [because it leads to] opportunism - more than 80% of those who take part are not really producers… I apologise for what I am going to say, but if I could I would drive over the settlers along the roads [waiting for land in agrarian reform projects]. These are not farmers, but criminals.

Along these lines, the most representative experience of entrepreneurial small family farming in the region today is the commercialisation of production by the cooperative Coopernova (a branch of the aforementioned Coopercana, which obtained autonomy in 1987). The great majority of the more than 2,000 associates are involved in numerous agrarian reform projects in several municipalities. Due to a combination of market opportunities and a fierce business mentality, the directors of the cooperative promoted the expansion and aggressive market insertion of agro-industrial production (especially dairy and processed food). When we travelled around the region, it was easy to identify family farmers supplying milk to Coopernova; these farms make more use of technology and irrigation, and also sell fruit. On the one hand, these farmers claim that milk production, despite the routine of daily hard work, is an attractive activity because of the constant flow of cash (one operational challenge is to maintain production during the dry season to avoid price reductions

---

[6] To a large extent, this replicates processes registered in the post-war years by Cândido (1964) in small farming areas of São Paulo, then occupied by a combination of indigenous and European migrant groups and increasingly influenced by the consumption of industrialised goods and the need to work outside the family plot. Ironically, the hard life of most families studied by Cândido led them to aspire to a better situation, "re-enacting conditions suggested in retrospective utopias" in Paraná, São Paulo and "even in Mato Grosso" (p. 157).

during the rainy season). On the other hand, the strong business orientation of the cooperative is often criticised by the associated farmers on the grounds that it has eroded previous partnership relations and dialogue with cooperative directors. New sources of income are now available, but many feel that this has led to the alienation of cooperative members and over-exploitation of labour in order to meet the quality and quantity demands of the more entrepreneurial cooperative. This was expressed in the following interview with a Coopernova associate:

> It is a lot of hard work with the cows, every day, you need to have a lot of determination, otherwise you don't stay in the game. I used to be a lorry driver, but I am trying my best to remain here. Milk production is not easy, and to maintain the farm one has lots of expenses, and annoyance. Sometimes I think about leaving, returning to my lorry [his previous job].

The pressure for a more entrepreneurial family agriculture has directly affected the perspective of colonisation and agrarian reform projects.[7] The first Regional Agrarian Reform Plan of Mato Grosso, covering the period 1986–1996, already aimed to reduce pressure for land through market-oriented approaches. An emblematic consequence was the 'market-friendly agrarian reform' [*reforma agrária de mercado*] project, launched in 1997 with World Bank support. Through loans to the Brazilian government, land was acquired and transferred (sold) to small family farmers. The objective was to streamline land distribution and reduce populism and transaction abuses. One emblematic example was the agrarian reform in Gleba Mercedes[8] in the Sinop region, involving 35,000 ha of (relatively large) 70 ha plots and around 500 families organised in nine rural nucleuses (Peripolli 2017), illustrated by a peasant farmer who migrated from the southern states to Mato Grosso (Fig. 4.2). Despite the high expectations for such an'alternative' model of agrarian reform, the project was seriously undermined from the outset, with accusations and evidence of inappropriate selection practices when determining who should receive land tenure; this was done through a lottery system that permitted applications from urban dwellers and the general public with no connection to agriculture. As a result, after the wood was cut and sold to the sawmills, many landowners sold their plots and failed to use the land for agricultural production. Illegal fires were also common in the agrarian settlements of Gleba Mercedes, which required special control by the fire brigade. In recent years, more land has been rented out to more capitalised residents and agribusiness farmers, especially for the cultivation of soybean. All those factors have significantly contributed to the negative image of farmers associated with the Gleba Mercedes project among the urban population and the agribusiness sector.

Once divested of the more political agenda, as in the case of 'market-friendly agrarian reform', PFF has been reinvented as the extension of agribusiness, playing the subordinate role of food supplier, labour reserve and land reserve. One important

---

[7]Data obtained from INCRA for December 2017 indicated the existence of 548 agrarian reform projects in MT, involving 82,424 families and covering 6,023,371 ha.

[8]This was part of the original 500,000 ha requested by Mercedes-Benz in 1968 in the context of the scramble for the Amazon.

Fig. 4.2 Peasant farmer of Gleba Mercedes

institutional change that has demobilised peasants and aggravated land commodification has been the concession of private land titles by the government (which happened in Gleba Mercedes in 2017), celebrated by mainstream politicians as proof of autonomy and an important step towards easier rural credit. Although some peasant families expressed satisfaction with the regularisation of their private property, they also acknowledged in the interviews that it will certainly lead to mass selling and concentration of land in the hands of wealthier residents and agribusiness farmers. Another important connection between the commodification of land in areas like Gleba Mercedes is the commodification of water for hydroelectricity, urban supply and irrigation (several large hydropower schemes are being constructed along the Teles Pires River—see Fig. 4.3). What is happening in Mato Grosso is part of national policies in favour of agribusiness expansion, suppression of land demands by peasants and indigenous groups, reduction of environmental conservation constraints and the renewed advance of extractivism and infrastructure projects.[9] One major knockback was the publication of a legal decree in 2016 (MP 759/2016 [known as the *MP da Grilagem* (Provisional Land Grabbing Decree)], converted into Law 13,465/2017, and further regulated by three additional decrees signed on 15 March 2018).[10] All

[9]The landslide victory of right-wing politicians in the 2018 general election is a disturbing indication of the power of agribusiness and its ideological, economic and political influence.

[10]Law 13,465 was challenged in the Supreme Court by the Public Prosecutor (ADI 5,771).

Fig. 4.3 Teles Pires River increasingly used for irrigation and hydroelectricity generation

this has facilitated the regularisation of public land occupied by large-scale owners and streamlined the transfer of property deeds to the farmers, invigorating the land market and facilitating socio-economic concentration in the hands of agribusiness players.

4.5 The Reluctant Alterity of Agribusiness at the Agricultural Frontier

What interests us most of all in this section is to reflect upon the ontological differences and continuities between present-day PFF and agriculture practiced for the market (i.e. agribusiness). From the brief overview above, it is possible to discern the internal complexity of contemporary PFF and the politico-economic pressures to reduce it to capitalist agriculture in agricultural frontier areas. There is an enduring 'friction' (Tsing 2005) between economic and social universalisms that clash with, and ultimately reconstruct, local cultures and idiosyncratic processes. The agricultural frontier is a dynamic social space where family farming is reorganised on the ruins of migrants' past experiences, combined with the practices of the original indigenous inhabitants and multiple demands for modernisation. Specifically in relation to our case study, after an initial phase in the 1970s and early 1980s, which was heavily reliant on the state apparatus, there has been a drive to incorporate the products and the operation of PFF into an agribusiness-centred regional economy, particularly in Mato Grosso, which has become the showcase of Brazilian agribusiness. The language and rationalisation of social groups involved in this transition to a market-based economy incorporate pre-capitalist beliefs and practices, as well as new forms of transgression and emerging, even disturbing, moralities (Taussig 1980). One very intriguing feature of this evolving geography is that small family farming certainly represents the other of capitalist agriculture, but it functions as a hesitant form of alterity that both resists and fulfils agribusiness.

The politico-economic literature can be of particular assistance, if tempered with sensitivity to frontier making. First of all, there are certainly no clear-cut boundaries between the agribusiness model of production and many of the strategies adopted by peasant farmers. This is what Fraser et al. (2017) consider a dialectics between mutuality/use values and markets/exchange values. Nonetheless, due to the multifaceted penetration of capitalism and the great diversity of peasants, the same literature amplifies significant confusion about the multiple features of PFF. For instance, Wolf (1966: 36) differentiates between peasants and entrepreneurial farmers, but also treats intensive small farmers in Western countries as 'neotechnic peasants' with an agriculture that is 'rationalized and transformed into an economic enterprise which aim[s] primarily at maximal outputs.' To avoid conceptual misunderstandings, it is useful to emphasise that the most crucial attribute of small family farming is not really the size of production, the family nucleus or the farming techniques, but the politicised relations between peasants and the expanding capitalisation of agriculture. This separation has taken place in concrete historical, spatial and political contexts, leading to the 'genesis of the capitalist farmer' due to the combination of new technologies, favourable supports, the usurpation of common land and also the impoverishment of 'the mass of the agricultural folk' (Marx 1981: 905–906).

When Marx situates the encroachment of capitalist relations on agriculture in time and space, he provides an important distinction to help us understand the role of small family agriculture. Marx highlights the specific relations of production and exchange that arise from the investment of capital in land, clearly distinguishing between peasant agriculture, which is pursued as livelihood, and capitalist agriculture, focused on profit making. Landed property is 'transformed by the intervention of capital and the capitalist mode of production… The assumption that the capitalist mode of production has taken control of agriculture implies also that it dominates all spheres of production and bourgeois society' (Marx 1981: 751). Nonetheless, the growing pressures for industrialisation and by large-scale agriculture do not mean the disappearance of 'small-scale peasant ownership, which continues to play an important role in modern economies' (Marx 1981: 942). But this sector has unique socio-economic features, because the peasant is the proprietor of their land and most production is consumed by the producers rather than traded as a commodity. There is a continuum between, at one extreme, 'traditional' subsistence farming and, at the other, highly capitalised, digital- and biotechnology-based agribusiness production. Crucially, at a certain point on this spectrum there is a qualitative transformation from peasant to agribusiness farming, which can be described as the effective and functional subordination and subsumption of peasant labour to capital. As theorised by Marx, subsumption is an effort to annul the political, economic and socio-cultural uniqueness of PFF, although the process is not absolute and many discontinuities remain (see Wanderley 2003).

Subsumption is a concept that Marx originally developed to explain the control of labour by capital during the expansion and consolidation of capitalist relations of production, but which has great applicability to explain the transformation of agriculture-cum-food into agriculture-cum-agribusiness at an economic frontier. The argument was developed in notebooks written in 1863-1866 and published as an appendix to Volume 1 of Capital. Basically, the labour process is subsumed under capital and the process of production becomes 'the process of capital itself', as in the case of the peasant who is employed as a day labourer for a farmer (Marx 1976: 1020). This is the stage of formal subsumption, when capital incorporates and gradually intensifies existing labour practices to increase the extraction of surplus value. Marx notes that high levels of subsumption do not necessarily mean that the life of the worker immediately deteriorates, but life is more dynamic and complex, as much as the labourers are more versatile and responsive. In his words: 'What a gulf there is between the proud yeomanry of England of which Shakespeare speaks and the English agricultural labourer!' (p. 1033). Interestingly it is quite revealing that Marx includes an old and intriguing quotation, which shows that the author did not neglect changes in agriculture and farming when thinking about subsumption: '*Agriculture from subsistence*... changed for *agriculture for trade*...' (Young 1774, in Marx 1976: 1035). This means that under pressures for the commercialisation of agriculture and commodification of nature, there is a gradual transformation of the labour process and of its actual conditions, a 'complete revolution' in the mode of production, which leads to the real subsumption of labour under capital.

The most relevant implication for our purposes here is that agricultural frontiers, as in the Amazon, because of fragile institutions, ideological pressures and the persistent subordination to core economic areas, are spaces where subsumption can more easily flourish (even though, due to the availability of natural resources peasants can also find escape routes and withdraw from market transactions more easily). Particularly in the eastern and southern parts of the Brazilian Amazon, land has been legally and illegally appropriated by companies controlled by international financial capital, with the enclosure of common areas, expropriation of local communities, the intensification of market pressures, increased conflicts and environmental damage (Frederico 2018). Yet, it is also pertinent to observe that in most cases the subsumption of PFF doesn't imply the separation of the producer from the means of production, in other words, it doesn't mean a full proletarianisation of peasants, but mainly their insertion into the sphere of capitalist relations as peasants. Subsumption happens because the balance between the internal characteristics of family farming and the external influence of capitalist agriculture has crossed a threshold and is disrupted in favour of capitalist relations of production. It is only at this stage that production for production's sake is accomplished and PFF begins to fade, subsumed under agribusiness. This process is certainly not new; by the early twentieth century, Chayanov (1991: 7) had already stated that it turns the autonomy of small agriculture producers 'into an economic system controlled on capitalist principles by a number of very large enterprises, which in turn are under the control of the highest forms of finance capitalism.'

Despite mounting controversies and class-based inequalities, the subsumption of PFF to capitalist agriculture has been advocated by numerous authors as something not only inevitable but highly advantageous to wider society in terms of economic intensification and rural development. Mendras (1984) famously announced that, under the inexorable pressures of industrial civilisation in the Western world, peasantry (as a unique way of life) ceased to exist and what remained were peasant producers (i.e. entrepreneurial family farmers), subordinated to market rules and the technological rigours of production. The period between 1945 and 1975, particularly in France, was called the 'Thirty Glorious Years' when people were tacitly excluded from agriculture due to increasingly automated production and additional management schemes. The word 'farmer' even replaced 'peasant', which was considered a rather obsolete term (however, more recently the word 'peasant' has been rehabilitated and again has more positive connotations). According to this line of reasoning, it is because of their partial connection with markets and a high degree of imperfection that PFF struggles to survive in capitalist societies (Ellis 1988). Likewise, Mellor (2017) sees a great future for 'small entrepreneurial farmers' (not subsistence peasants) in terms of poverty reduction and rural development, while Graeub et al. (2016) make a distinction between family farmers who are well-endowed and well-integrated into markets and those who lack sufficient credit or who are land-poor and require social safety nets. Abramovay (1990), among others Brazilian scholars, has applauded the transition from PFF to entrepreneurial family farming, under the belief that these farmers will be able to benefit from technical innovation and flourish in the free market economy (illustrated by the local food market in Sinop—Fig. 4.4). This pro-subsumption position was influenced by the apparent success of the small units of production boosted by dedicated state interventions in the European Union and the Brazilian peasants who Abramovay (1998) considers ruined and backward.

The subsumption of PFF to agribusiness has been encouraged in Brazil not only because of market pressures but also through the very official programmes designed to support peasant farming. PRONAF, in particular, became a mark of the Lula government (2003-2010) and introduced a series of incentives focused on preferential commercialisation, formation of strategic stocks, seed provision, etc. (Cazella et al. 2009). The related Food Purchase Program (PAA) guaranteed priority markets for the products of small family farmers in an attempt to stimulate the local economy and improve food security. The scheme represented a great boost to some segments of family farmers because the legislation stipulated that 30% of food purchase was reserved, although in practice this never reached more than 15% because it was often difficult for peasant farmers to respond (B. Fernandes, UNESP, pers. comm.). In spite of its ambitious goals, the PAA programme suffered from a series of operational problems and significantly declined after 2013 because of the growing scarcity of public funds; in any case, PAA only benefited 4.2% of family farmers in Brazil (Grisa and Porto, 2015). Closely related to PAA was the National Programme of School Food (PNAE), which also secured a share of the market for small family farmers

Fig. 4.4 Fruit and vegetable market of Sinop

who supplied the municipal school system. Schemes like these are described in the literature as 'institutional markets', given that they represent an attempt by the state to control food trade and supply (it should be noted that all markets are in effect institutionalised and follow social norms and customs).

Regardless of the enthusiasm of many, these so-called institutionalised markets are circumstantial attempts to limit the perverse influence of agri-food corporations and their supply chains rather than an effective mechanism to contain agribusiness pressures. A related weakness is that government initiatives that benefit the poor strongly depend on political backing and can quickly be eroded with government changes (exactly what happened in Brazil in 2016 after the removal of the left-wing president Dilma by the National Congress). As a result, the softened path of subsumption associated with institutionalised markets is likely to be overtaken by the stronger pressures exerted by the agribusiness sector, which tends to aggravate inequalities and maximise risks.

4.6 Wrapping Up: Disconcerting Prospects

Agricultural frontiers in the Amazon are hotspots of agrarian transformation and intensely reveal old and new development trends. Significant contingents of family farmers have moved to Mato Grosso and other parts of the region since the 1960s and taken part in the organisation of a new peasantry that has replaced, through various forms of violence, the previous peasants and indigenous peoples. A significant proportion of those who migrated to the region lost their land and either returned to their places of origin, moved to urban centres or were proletarianised in agribusiness farms. Among those who succeeded, new challenges appeared due to the affirmation of the power of agribusiness and the exponential growth of soybean, cotton and maize production, together with irrigated crops and agroindustry. Overall, this represented a mounting pressure for the subsumption of PFF to the rationality of agribusiness, only circumstantially mitigated by various government programmes (which at any rate operate with similar pro-subsumption goals). The absence of a fully developed capitalist mode of production, as observed by Marx (1976), did not preclude the subsumption and exploitation of peasant labour, particularly in these areas of agricultural frontier where both capitalist and non-capitalist relations are in a state of flux. These circumstances have really evolved through a double process of subsumption: the agricultural frontier being subsumed to the hegemonic sociopolitical and economic influence of central areas, with PFF simultaneously being subsumed to capitalist relations across scales.

It is important to realise that subsumption to market relations has been instrumental for the very success of export-oriented agribusiness. The mounting risks associated with agribusiness farming, increasingly affected by environmental degradation, pesticide-related diseases and labour scarcity, need escape routes that are obtainable in the universe of PFF around large-scale properties. There are cases, such as the agrarian reform community Zumbi do Palmares in the north of Mato Grosso, where PFF not only produces significant amounts of food but also guarantees the maintenance of vicinal roads and public schools for families living on local farms. The pressure of subsumption is furthermore an attempt to deny small farmers the prospect and the possibility of resistance and reaction to market demands. What is meant by resistance includes a variety of strategies, ranging from moments of open confrontation (as in the case of the occupation of land by landless groups) to the more common acts of passive defiance and the formation of location-based networks around kinship and friendship (see Scott 1985). The subsumption of PFF to the requirements of agribusiness is never complete, but meets the opposition of peasant farmers, who retain the ability to adapt, are highly diversified (in terms of production and activities) and know better how to cope with local conditions and natural cycles of production and develop strategies that make little sense to capitalist players (over-exploitation, engaging the family, relations of loyalty and retribution, etc.).

The intricate Brazilian experience of approximation and distancing from agribusiness has parallels with what has been happening with peasants all over the world. Contradicting the pessimistic predictions of the twentieth century, there has been a revival of peasant struggles for social emancipation and legitimate rights of access to land and food in the context of mounting domination of financial capital and neoliberalism (Herrera and Chi 2018). Small farmers maintain an 'inconvenient' and resilient autonomy that is not at all easy to erode; they exhibit an internal logic (beyond romanticism) and flexibility that is difficult for the reductionist operations of capitalism to incorporate. The behaviour of peasants in response to onerous socio-economic relations follows normative attitudes that Thompson (1993: 188) famously described as a 'moral economy', which is not political 'in any advanced sense, nevertheless, it cannot be described as unpolitical either, since it supposed definite, and passionately held, notions of commonweal.' That is, family farming, through various survival strategies (beyond the pressures to expand and modernise) is not a residual or marginal category of farm operations left behind by modernisation, but a conscious economic strategy and approach to rural life (Machum 2005). Still, the interplay between survival, resistance and opportunities are even more intense in areas like the ones in the Amazon studied here, where agrarian questions and the ability of peasants to respond to challenges are directly related to the range of inequalities and alliances that result from the specific circumstances of frontier making.

References

Abramovay R (1990) De Camponeses a Agricultures: Paradigmas do Capitalismo Agrário em Questão. PhD thesis, Unicamp, Campinas
Abramovay R (1998) Paradigmas do Capitalismo Agrário em Questão, 2nd edn. São Paulo/Campinas, Hucitec/Unicamp
Adams C, Murrieta R, Neves W, Harris M (eds) (2009) Amazon peasant societies in a changing environment. Springer
Arulbalachandran D, Mullainathan L, Latha S (2013) Food Security and Sustainable Agriculture. In: Arulbalachandran D (ed) Sustainable agriculture towards food security. Springer, Singapore, pp 3–13
Asturias MA (2005[1949]) Hombres de Maíz. Madrid, Alianza Editorial
Barta A (2011) Os Novos Camponeses (trans: Pandolfi MA). UNESP/UNESCO/Cultura Acadêmica, São Paulo
Bernstein H, Friedmann H, van der Ploeg JD, Shanin T, White B (2018) Forum: fifty years of debate on peasantries, 1966–2016. J Peasant Stud 45(4):689–714
Brondizio E (2009) Agricultural intensification, economic identity, and shared invisibility in Amazonian peasantry: Caboclos and colonists in comparative perspective. In: Adams C, Murrieta R, Neves W, Harris M (eds) Amazon peasant societies in a changing environment, pp. 181–214. Springer
Brookfield H, Parsons H (2007) Family farms: survival and prospect. Routledge, London and New York
Caldeira J (2017) História da Riqueza no Brasil. Estação Brasil, Rio de Janeiro
Campbell JM (2015) Conjuring property. University of Washington Press, Seattle and London
Cândido A (1964) Os Parceiros do Rio Bonito. José Olympio, Rio de Janeiro
Carolan M (2013) Reclaiming food security. Routledge, London and New York
Cazella AA, Bonnal P, Maluf RS (eds) (2009) Agricultura Familiar: Multifuncionalidade e Desenvolvimento Territorial no Brasil. Mauad X, Rio de Janeiro
Chayanov A (1966) The theory of peasant economy (eds: Thorner D, Kerblay B, Smith REF). American Economic Association, Homewood, IL
Chayanov A (1991[1927]) The theory of peasant co-operatives (trans: Benn DW). Ohio State University Press, Columbus
Clapp J (2018) Mega-mergers on the menu: corporate concentration and the politics of sustainability in the global food system. Glob Environ Polit 18(2):12–33

Clapp J, Fuchs D (eds) (2009) Corporate power in global agrifood governance. MIT Press, London and Cambridge, MA
De Schutter O (2017) The political economy of food systems reform. Eur Rev Agric Econ 44(4):705–731
Desconsi C (2011) A Marcha dos Pequenos Proprietários Rurais: Trajetórias de Migrantes do Sul do Brasil para o Mato Grosso. E-papers, Rio de Janeiro
Ellis F (1988) Peasant economics: farm households and agrarian development. Cambridge University Press, Cambridge
Foweraker J (1981) The struggle for land: a political economy of the pioneer frontier in Brazil from 1930 to the present day. Cambridge University Press, Cambridge
Fraser JA, Cardoso T, Steward A, Parry L (2017) Amazonian peasant livelihood differentiation as mutuality-market dialectics. J Peasant Stud 45(7):1382–1409
Frederico S (2018) Território, Capital Financeiro e Agricultura. Livre-docência thesis. UNESP, Rio Claro
Friedmann H (1980) Household production and the national economy: concepts for the analysis of agrarian formations. J Peasant Stud 7(2):158–184
Garner E, O Campos AP (2014) Identifying the "family farm": an informal discussion of the concepts and definitions. ESA working paper no. 14–10. FAO, Rome
Graeub BE, Chappell MJ, Wittman H, Ledermann S, Kerr RB, Gemmill-Herren B (2016) The state of family farms in the world. World Dev 87:1–15
Grisa C, Porto SI (2015) Dez Anos de PAA: As Contribuições e os Desafios para o Desenvolvimento. In: Grisa C, Schneider S (eds) Políticas Públicas de Desenvolvimento Rural no Brasil. UFRGS, Porto Alegre, pp 155–180
Hébette J (2004) Cruzando Fronteira: 30 Anos de Estudo do Campesinato na Amazônia. Belém, UFPA
Herrera R, Chi LK (2018) The convergence of peasant struggles worldwide. Econ Polit Wkly 53:42–49
IAASTD (2009) Agriculture at a crossroads: a synthesis of the global and sub-global IAASTD reports. International Assessment of Agricultural Knowledge, Science and Technology for Development, Washington, DC
Ilbery B (ed) (1998) The geography of rural change. Pearson/Prentice Hall, Harlow
Ioris AAR (2012) The geography of multiple scarcities. Geoforum 43(3):612–622
Ioris AAR (2016) Rent of agribusiness in the Amazon: a case study from Mato Grosso. Land Use Policy 59:456–466
Ioris AAR (2017) Agribusiness and the neoliberal food system in Brazil: frontiers and fissures of agro-neoliberalism. Routledge, London
Ioris AAR (2018a) The politics of agribusiness and the business of sustainability. Sustainability 10(5):1648. https://doi.org/10.3390/su10051648
Ioris AAR (2018b) Amazon's dead ends: frontier-making the centre. Polit Geogr 65:98–106
Machum S (2005) The persistence of family farming in the wake of agribusiness: a new Brunswick, Canada case study. J Comp Fam Stud 36(3):377–390
Magdoff F, Tokar B (eds) (2010) Agriculture and food in crisis: conflict, resistance, and renewal. Monthly Review Press, New York
Marx K (1976[1867]) Capital, vol 1 (trans: Fowkes B). Penguin, London
Marx K (1981[1894]) Capital, vol 3 (trans: Fernbach D). Penguin, London
McDonald BL (2010) Food security. Polity, Cambridge
Medina G, Almeida C, Novaes E, Godar J, Pokorny B (2015) Development conditions for family farming: lessons from Brazil. World Dev 74:386–396
Mellor JW (2017) Agricultural development and economic transformation. Palgrave Macmillan
Mendras H (1984) La Fin des Paysans, 2nd edn. Paris, Actus Sud
Nugent SL (1993) Amazonian Cabloco society: an essay on invisibility and peasant economy. Berg, Berg

Nugent SL (2002) Whither O Campesinato? Historical peasantries of Brazilian Amazonia. J Peasant Stud 29(3–4):162–189
Otero G, Gürcan EC, Pechlaner G, Liberman G (2018) Food security, obesity, and inequality: measuring the risk of exposure to the neoliberal diet. J Agrar Chang 18:536–554
Peripolli OJ (2017) Formação Continuada – O Projeto: Por Entre Estradas e Trilhas, o Caminho que se Fez e se Faz ao Caminhar. In: Peripolli OJ, Neideck RMB, Zoia A (eds) Formação Continuada. Univates, Lajeado, pp 31–47
Queiroz MIP (1976) O Campesinato Brasileiro, 2nd ed. Vozes, Petrópolis
Redclift M (1997) Sustainability and theory: an agenda for action. In: Goodman D, Watts MJ (eds) Globalising food. Routledge, Abingdon, pp 333–343
Santos JVT (2008) Programa de Colonização Terranova. In: Barrozo JC (ed) Mato Grosso: Do Sonho à Utopia da Terra. EdUFMT/Carlini & Caniato, Cuiabá, pp 97–140
Scott JC (1985) Weapons of the weak. Yale University Press, New Haven and London
Scoville JM (2015) Framing Food Justice. In: Dieterle JM (ed) Just food: philosophy, justice and food. Rowman & Littlefield, London and New York, pp 3–20
Shanin T (ed) (1987) Peasants and Peasant societies, 2nd edn. Oxford, Basil Blackwell
Shanin T (2006) Peasantry. In: Outhwaite W (ed) The Blackwell dictionary of modern social thought, 2nd edn. Blackwell, Oxford, pp 467–469
Shiva V (2016) Who really feeds the world? The failures of agribusiness and the promise of agroecology. North Atlantic Books, Berkeley
Taussig MT (1980) The devil and commodity fetishism in South America. University of North Caroline Press, Chapel Hill
Thompson EP (1993) Customs in common: studies in traditional popular culture. New Press, New York
Tsing AL (2005) Friction: an ethnography of global connection. Princeton University Press, Princeton and Oxford
UNICEF (2017) The state of the world's children 2017: children in a digital world. UNICEF, New York
Van Vliet JA, Schut AGT, Reidsma P, Descheemaeker K, Slingerland M, van de Ven GWJ, Giller KE (2015) De-mystifying family farming: features, diversity and trends across the globe. Glob Food Secur 5:11–18
Walford N (2003) Productivism is allegedly dead, long live productivism: evidence of continued productivist attitudes and decision making in South-East England. J Rural Stud 19:491–502
Wanderley MNB (2003) Agricultura Familiar e Campesinato: Rupturas e Continuidade. Estudos Sociedade e Agricultura 21:42–61
Wanderley MNB (2014) O Campesinato Brasileiro: Uma História de Resistência. Revista de Economia e Sociologia Rural 52(supl 1):S025–S044
Watts MJ (2013[1983]) Silent violence: food, famine, and peasantry in Northern Nigeria. University of Georgia Press, Athens and London
Wilson GA, Burton RJF (2015) 'Neo-productivist' agriculture: spatio-temporal versus structuralist perspectives. J Rural Stud 38:52–64
Wolf ER (1966) Peasants. Prentice-Hall, Englewood Cliffs, NJ

Chapter 5
Water and Energy Frontiers in the Amazon

Abstract The chapter enlarges the examination of the notion of 'resource frontiers' and examines how the construction of dams and other related water infrastructure, as it has been happening particularly on the Brazilian section of the Amazon, is time and again used to propel and celebrate a nation's modernisation, but it also serves to reveal how modernity is always partial, fraught with gaps and contradictions. The politics of modern water management in the Brazilian Amazon is scrutinised from a cultural, political and economic perspective that aims to contribute to an understanding of the politics of distribution, recognition and resignification that play an important role in the evolution of water modernity in the region. The chapter proposes a periodisation of regional water management, divided into three moments of intense socioecological change and political and economic influences. In that regard, Cultural Political Economy (CPE) can facilitate the understanding of the impact of hegemonic pro-development ideologies and discourses, as well as to comprehend the complexity of the interface between economic and more than economic practices.

Keywords Dams · Navigation · Belo monte · Recognition · Resignification

5.1 Introduction

One of the main processes of change in the Amazon Region today, which closely follows the production of abundances at the expense of new forms of scarcity in the agricultural sector, is the construction of large-scale water projects, particularly for hydroelectricity generation and river navigation (associated with export-oriented agribusiness, timber harvesting and mining). Hydropower engineering at the frontiers of national development in Brazil is significantly strategic, among other reasons, because it supplies electric energy to the main urban and industrial sites, despite all the controversies and injustices that surround them. Water infrastructure projects have been built through various forms of public–private association (mostly reliant on public funds and encouragement from governmental institutions) and have invariably caused widespread social, cultural and ecological impacts. Those negative consequences of large water infrastructure projects reflect the interplay between the pressures for economic growth exerted from the main politico-economic centres

A. A. R. Ioris, *Frontier Making in the Amazon*, Key Challenges in Geography,
https://doi.org/10.1007/978-3-030-38524-8_5

and the unique geographical circumstances of the regional socio-spatial frontiers. Building on the material discussed earlier, this chapter will examine the association between water management, agricultural expansion and hydropower generation, and will consider the controversial water-agriculture-energy nexus in relation to the wider phenomenon of frontier making. Approaching these issues from a politico-economic perspective, the aim is to contribute to understanding the dialectics between what is unique and local with the more widespread and global features of water modernity. In that regard, the chapter will deal with what Bourdieu (2004) defines as the tension between history and reason. It will examine the three main themes: policy-making, economic activity and nature commodification, which broadly correspond to the periodisation of the agricultural frontier of Mato Grosso. This topic will be specifically analysed here from a cultural, political and economic perspective that aims to contribute to an understanding of the politics of distribution, recognition and resignification that play an important role in the making of socio-spatial frontiers.

Let's start our discussion with a graphic image that, although distant from the Amazon, has nonetheless a number of important and revealing parallels in terms of the production of modern socio-spatial frontiers. The scene is the Russian summer, in August 1933. A group of 120 leading writers went on a boat tour along the recently inaugurated White Sea Canal, one of the emblematic expressions of the supposed success of the first Five-Year Plan of the Soviet regime. The canal was built at break-neck speed between 1931 and 1933 using the manual labour of 120,000 workers, most of the prisoners were sentenced for minor political offences or, in many cases, having committed no crime whatsoever but rather falling prey to intrigue and anonymous denunciations. Prisoners were basically a source of slave labour—more than 10% died during the construction of the canal—and were sent to the construction site supposedly to reform themselves until they could be accepted and reintegrated back into society. Imprisonment was at that time justified as a humanitarian project to rehabilitate 'bad people', that is, in practice anyone who had an opinion about the corrupt and ultra-violent regime of comrade Stalin. However, the abundant evidence of cruelty and state propaganda associated with the White Sea Canal did not affect the mood during the festive trip. On the contrary, as vividly described by the historian Orlando Figes (2008), the guests were in high spirits, enjoying the best food and drinks (including caviar, cheese, wines, cognac and smoked sausage), oblivious to the famine and political persecution affecting most of the country, and discussing literature with top Politburo members, party functionaries and the 'great man' himself, Joseph Stalin.

After a reception in a luxurious hotel, the group went to inspect dams and locks and attended a theatrical presentation put on by prisoners (although they were not allowed to talk to the prisoners directly). The trip was considered very successful and, as a sign of gratitude, most of the writers contributed to a book published to celebrate the completion of the canal. This project, along with others such as the Moscow–Volga canal and Aral Sea irrigation canals, is an example of a water infrastructure project that cost lives and reputations and destroyed scarce resources for almost nothing, given that navigation along the canal was always limited due to construction mistakes that left the waterway too shallow to carry anything but the smallest boats.

However, the trip was highly beneficial for the purposes of a notoriously tyrannical government aiming to manipulate, control and exploit its people. The Five-Year plans (there were 13 in total) were clearly influenced by the modernist vision of Lenin, who introduced rigid and dictatorial politics after the victory of the October Revolution (contradicting what he himself had written about democracy and the state just a few weeks earlier) and launched technocratic administrative strategies loaded with scientificism, positivism and a striving for modernity (Scott 1998). Stalin dutifully followed Lenin's ideas of discipline, the indisputable wisdom and leadership of the vanguard party, the justification of extreme violence and the need to severely punish all challengers. Interestingly, and tragically, the monumental crimes committed by the Soviet state on behalf of the workers had already been denounced in the early days of the Revolution by Rosa Luxemburg, who believed that the party should lead but also learn from the masses, that is, she recognised the importance of respect for human experience and agency. For Luxemburg, the revolution was a living process that could not be enforced from above (her sensibilities and emphasis on sociocultural complexity will be relevant when we return below to the frontiers of water, agriculture and energy in the Amazon.).

The key point is that what happened in Russia in the middle of the last century anticipated or mirrored many other situations in which the control of water became closely associated with economic development goals, nationalist ideologies and the affirmation of modern values. These include the Marathon Dam in Greece, Nehru's dams in India (deviating from Gandhi's opinion about postcolonial development), Roosevelt's dams, navigation and irrigation along the Tennessee Valley as part of New Deal strategies, interconnected hydropower schemes in the Scottish Highlands and dams in the Upper Douro in Portugal, the Aswan Dam in Egypt, among many others. The story is well known and repetitive. The construction of dams and other related infrastructure is time and again used to propel and celebrate a nation's modernisation, but it also reveals the ontological impurity of modernity, always partial, fraught with gaps and contradictions. For instance, the contemporary model of water use and management, greatly influenced by the agenda of ecological modernisation and sustainability, comprises a problematic combination of alleged rationality and efficiency gains with the privatisation and degradation of water bodies (Ioris 2018). Considering this international experience, our goal here is not to revisit economic statistics or the details of engineering projects, but to assess the main direction and the internal contradictions of prevailing development trends. The lived spaces of the Amazonian frontier have been dramatically transformed by sustained migration, new production technologies, urban expansion and new relations of production. Growing towns, farms, industries and households all need electricity, which means that the construction of hydropower schemes is one of the top infrastructure priorities for national, regional and local policymakers and politicians. Water continues to largely define the Amazon and is increasingly a politico-economic issue.

5.2 Politics of Distribution, Recognition and Resignification

The cultural political economy of water, situated in the wider context of socio-spatial frontiers, is a privileged entry point into the contemporary problems of development, ecological conservation and environmental justice. Cultural political economy (henceforth, CPE) emerged in the 1990s as part of the 'cultural turn' advocated by social science scholars against the historicist or transhistorical analyses that had so far characterised disciplines such as sociology, economics and human geography. It was a concerted attempt to take seriously the complex relations between meanings and practices that underpin economic relations (Jessop and Oosterlynck 2008). The strength of CPE has principally been its ability to make sense of capitalism's hegemony and expansionism without reducing all analysis to the economic sphere. This means recognising that there are spaces and interstices in human relationships, notably in the practices of daily life, family traditions and aspirations, that flourish and proliferate beyond the economy. Work on CPE is concerned with the variety and non-linearity of lived economic experiences, what Mann (2012) calls the 'sidewalk path' dimension that complements the 'main flow' on the street. Culture-sensitive political economy should, therefore, connect the economic and non-economic elements of world complexity. Two influential authors in this debate, Jessop and Sum (2006), combined Gramscian and Foucauldian ideas to explore the relationships between economic imaginaries and intersubjective meanings, arguing that CPE occupies a middle ground between reified economics (the naturalisation of economic categories) and sociological approaches that address sociocultural activities and micro-level interactions.

The cultural turn, an academic movement beginning in the early 1970s inspired by postmodern and post-structuralist theories, had already highlighted the importance and substantive influence of discourse, feelings and representations reflected in the functioning and the contradictions of modernity and capitalism. Its scholars challenged reified economic structures and the pre-assumed priority of economic processes above other sociocultural phenomena, paying renewed attention to experienced, micro and qualitative dimensions of human activity. One central aspect of this debate is how some imaginaries and paradigms are selected in a particular sociocultural conjuncture to inform policies and state interventions (Sum and Jessop 2013). Closely associated with the cultural turn, CPE demonstrated the convergence of discursive, structural, technological and agential mechanisms. CPE has helped to restore the weight of identities, discourses, work cultures and the social and cultural embedding of economic action (Sayer 2001). Having said that, however, work on CPE has also reflected the excesses and exaggerations common to post-structuralist approaches, in particular, the reductionist tendency to privilege the immaterial and symbolic dimensions of reality. Post-structuralist authors typically emphasise the primacy of flat ontologies and specific circumstances, but at the expense of generalisations and commonalities that also need to be recognised. If it is an important achievement to denounce the narrow economic thinking behind the conventional political economy, it is unwise to disregard the strong influence of economic forces, market transactions and institutional settings. And we probably do not require grand narratives to recognise, as pointed out by Sayer (1992: 4), that there is not only difference but also necessity in the world (although 'we must think [of] necessity as the becoming-necessary of the encounter of contingencies', according to Althusser 2006: 261).

In order to determine its positive attributes and avoid focusing too much on culture and not enough on the economic and political dimensions, it is necessary to situate the cultural political economy in the long trajectory of the political economy. The evolution of the political economy, as an interdisciplinary field of research that appeared in the eighteenth century, reveals a great deal about the challenges involved in interpreting capitalist relations of production and reproduction. The rapid expansion of industry and of urban centres of consumption, as well as their connections with agriculture and with sources of raw material, increasingly required novel explanatory methods. Informed by the Enlightenment's rationalist and materialist thinking, 'classical' political economy (also described as bourgeois economics) advanced social theories on the basis of society's economic foundations and market forces. These theories, most famously in the case of Adam Smith, dealt with analytical categories such as value, property, production, market rules, classes and the state. Political economy also closely followed Lockean liberalism and the centrality of private property for the maintenance of liberty, privacy and freedom. Beyond the economist focus adopted in Britain, Hegel (2008: 187) argued that the political economy is 'one of the sciences which have arisen out of the conditions of the modern world', and it aims to explain the complexity, the qualitative and quantitative character of mass-relationships and mass-movements.

In clear opposition to the initial liberal and bourgeois perspective, Marx reworked the contribution of earlier authors, particularly Ricardo's value theory, to describe capitalism's reliance on the exploitation of workers and the commodification of labour. Marx (1976: 174) denounced the failure of classical political economy to recognise the role played by abstract labour, that is, the fundamental transformation of value into exchangevalue in a capitalist economy. In his opinion, the classical political economy is only able to view the economic question in relation to manufacturing, the accumulation of capital, and maximising commodity production and the extraction of profit from labour. As a result, the world of commodities and the symbolic power of money conceal 'the social character of private labour and the social relations between the individual workers, by making those relations appear as relations between material objects, instead of revealing them plainly' (Marx 1976: 168–169). More than 20 years before the publication of Capital, Marx had already put forward a humanist critique of the unfairness and perversity of economic trends, including the claim that 'material, immediately sensuous private property is the material sensuous expression of estranged human life' (Marx 1988: 103). Around the same time, Marx attacked the illusionary results of conventional analyses that separate deeply connected historical processes, focusing only on princes, states and religion, and leading to fragmentation and mystification. Not only are sociocultural issues disregarded and dismissed, but with this 'the relation of man to nature is excluded from history and hence the antithesis of nature and history [i.e. culture] is created' (Marx and Engels 1974: 59).

However, the most influential critique of classical political economy in the Western world certainly did not come from Marx, but in the form of marginalist economics, which emerged and expanded between the 1860s and the 1890s. Marginalism was independently introduced by Jevons (England), Walras (Switzerland) and Menger (Austria), and its primary focus was the determination of prices through consideration of the marginal utility of goods and services, rather than focusing on value, order and economic growth as in political economy. Marginalist economics centred on individualism and subjectivism (instead of grand interpretations of economic trends or the discussion of fundamental analytical categories), arguing that value is determined by the consumer's utility. The 'marginalist revolution in economics', which incidentally has significant relevance for today's neoliberal policies, was considered a great improvement among conservative circles as it freed the political economy from rigid categories and helped to found modern, 'scientific' economics. For instance, Marshall's 1890 marginalist elaboration on public amenities and his marginal theory of value later inspired Pigou, in 1920, to describe environmental problems as a divergence between 'marginal social net product' and 'marginal private net product'. According to the marginalist perspective, those that benefit from the use of the environment should internalise the social costs (externalities) of their activities via, for example, the payment of fees and taxes. Also Coase, in 1960, understood that government intervention was less important for ensuring the adequate use of resources since bargaining between players constituted a more effective solution; this means that, as long as a regime of explicit ownership can be established, water allocation and pollution problems are solved rationally, as long as water can be bought and sold through the market (Ioris 2010).

Marginalism was the final step in the removal of political economy from mainstream scholarship and its containment primarily within the texts of Marxist authors (Clarke 1991). This is not the place to revisit the continuities and antagonisms between Marxist and marginalist (neoclassical) economics, but at least some crucial observations are in order to avoid the risk of CPE being captured by marginalist ideas. At face value, it may seem logical that due to CPE's sensitivity to culture, individual agency and subjectivities, it could be more comfortably associated with the reasoning of marginalist economics. In addition, both CPE and marginalism could be seen together in the same attempt to overcome the structuralist, all-embracing biases of classical political economics. Both attack the apparent narrowness of many political economy studies, which struggle to deal with the intricacies of the world due to a fixation on economic processes and, as a result, provide limited, overambitious and simplistic interpretations. However, beware! Here lies a trap, the 'marginalist trap': despite its methodological emphasis on individual preferences and people's specific circumstances, marginalist economics is, in fact, built on a clear positivistic paradigm that proclaims the superior rationality of market freedoms in order to impose business rationality on all dimensions of human experience. Standing in clear contrast with the gains achieved during the cultural turn, marginalism is actually even more rigid, ideological and schematic that the politico-economic argument that it attempted to subvert, particularly because of its reduction of the complexity and non-linearity of the world to the monodimensionality, alienation and utilitarism of market transactions (commonly mystified through the uncritical use of statistics and mathematical models). In other words, the prominence of subjectivism and individualism did not prevent marginalist thinking from dissociating economic processes from their sociocultural context.

However, it is not enough only to denounce to merely avoid the marginalist trap of marginalism bait; it is also important to retain maintain CPE firmly within the politico-economic critique of disembedded markets and of capitalist relations of production and reproduction. If CPE is a great beneficiary of the cultural turn, it is possible to argue that the emphasis on subjectivities, differences and uniqueness may have gone too far, at least when post-structuralist authors deny the importance of generalised trends or of the lived impacts of structural and functional characteristics of social relations. Commonly, exaggerated prominence is given to flat ontologies and non-hierarchical relations at the expense of conscientious and nuanced economic and sociological thinking. This must serve as an alert to the overindulgences of the cultural turn and the danger of eroding political and economic dimensions in favour of a messy and shallow 'cultural bazar'. This is because the illuminating focus on the social and cultural embedding of economic activities cannot merely ignore the contribution of political economists to unlocking the complexities of capitalist socioeconomy (Sayer 2001). The good news is that there is ample space within the political economy itself to put into practice any cultural, phenomenological and semiotic sensibility. In addition, despite the fact that in the public imaginary political economic approaches have practically become a surrogate for vulgar Marxism and its grotesque practical renderings (such as the absurdities of Stalinist planning), it is worth salvaging the humanist kernel of Marx's politico-economic contribution. One of Marx's main goals, in Capital and other texts, was to go beyond the mysterious, enigmatic basis of commodification and consider the associated relations of production and social control. The 'magic' of the commodity is something that belongs to bourgeois economics and needed to be demystified through a critique of political economy.

It would be anachronistic and misleading to reinvent Marx as a champion of post-structuralism, because he was really a child of the Victorian era, fuelled by optimism about technological developments and scientific innovation. Moreover, Marx was able to brilliantly rework and radically reconnect economics with sociopolitical processes and some implicit attention to culture and diversity. Marx was no determinist, but he understood that history was wide open, so that many possibilities exist at any point in time and key human activity is not reduced to the economic (although the economic was obviously hugely important to Marx, as pointed out by Eagleton 2011). It is not difficult to perceive here that Marx was an acute critic of the social destabilisation produced by the advance of commodity production, and he mobilised political economy in a plan to radically challenge the capitalist world order. For Marx (1976), different products are equated as values through market exchange, as different kinds of labour are equated and standardised (abstracting the 'real inequality'). The consequence is that the product of labour is transformed into a 'social hieroglyphic' in which specificity and concreteness are dissolved. This phenomenon, according to Marx (1976: 167), is 'as much men's social product as is their language'. There is, thus, a humanist and non-economic basis to CPE that can be traced back to Marx's analysis and revitalisation of classical political economy, particularly through his detailed investigation of the role of labour in capitalist production and his focus on the relations and consciousness that produce the material reality of the world. In the end, Marx's political economy is permeated by his philosophical humanism, his philosophy of freedom and self-understanding (Jeannot 2010).

Another important contribution of Marx's work was his analytical method of connecting the parts with the whole, relating politico-economic expression with the historical structure that enables and constrains. There is, therefore, a dialectical unity linking historical and economic trends with territorialised and culturally specific processes; in other words, the dialectics between history and sociocultural meaning. Marx strived to go beyond simplification when dealing with the contingent, historical and cultural features of specific capitalist societies, as in the case of the United States, about which he affirmed that the 'perfected Christian state is rather the atheist state', while the 'so-called Christian state is the Christian negation of the state, but it is certainly not the political realisation of Christianity' (Marx 1975[1843]: 222–223). Marx always articulated his ground-breaking analysis through an inquiry into the category of totality, that is, the pursuit of a logical construct that refers to the way the whole is present through internal relations in each of its parts (Ollman 2003). This is even more relevant considering that, as pointed out by Bourdieu (2004), objectivity and truth are intertwined and observe localised sociocultural constructs that nonetheless cannot merely be reduced to the social conditions and constructions that made them possible. Therefore, if the humanistic and cultural sensitivity of Marxist political economy and its creative methodology are taken on board, the culturalistic excesses of CPE (i.e. an excessive focus on linguisticality, textuality or symbolism at the expense of the material and socioecological context) might be turned on their head. Rather than resulting in an abundance of ideas and creativity, the rehabilitation of humanist and ecological materialism leads to the intriguing conclusion that CPE may have not gone far enough.

It has been a major achievement to instigate the integration of culture and identity into the work of conventional political economy, but what is normally missing in these approaches is the wider context of society and the economy, that is, the interconnections and interdependencies between society and the rest of nature. The desired recognition of the importance of culture, however, needs to be closely associated with a fundamental concern for justice, not only locally, but nationally and globally. As argued by Badiou (2012), it is crucial to understand the diversity of the world, which essentially means understanding its regimes of relations of identity and difference, but the main philosophical and political task today is the search for justice; if the aim is to get rid of the structural and perverse ramifications of Western dominance, there is no time to waste on relativisms (for Badiou, the search for justice is a concrete expression of political truths that subverts the widespread relativism of postmodern thought). Geographers in, particular, have called for a new 'turn', able to emphasise materialist sensibilities and the transformative work of meaning-making cultural processes in the concreteness of the world (Kirsch 2012). Likewise, the debate about justice needs to reject the economistic view that reduces the question of recognition to a mere epiphenomenon of distribution; in effect, the two—recognition and distribution—are expressions of different moments, but they need to be articulated and theorised together (Fraser and Honneth 2003).

It is highly significant for the revitalisation of the politico-economic argument that Sayer (2001: 695) advocated a CPE based on a fuzzy distinction between the system (formal structures beyond the subjective experience of actors) and lifeworld (related to experience, informality and the culture in which actors socialise) that cuts across the "culture-economy" and "recognition-distribution" binaries. The specific prominence of recognition complements the conventional emphasis of political economy on the question of redistribution, that is, CPE needs to restore the balance between the multiple and intricate dimensions of lived reality. Recognition, including multiple struggles to assert identity and difference, is a central concept in the world today, in a context of accelerated transcultural contact, whilst distribution has historically been associated with the claims of the working class and the poor during previous phases of capitalist history. Nonetheless, it is not sufficient to pay lip service to the removal of binaries; an accomplished dialectical reconstruction of CPE requires extending the politics of distribution and recognition, with the addition of the politics of resignification. This third concept, resignification, is increasingly needed to account for the interdependencies between the economic and more than economic manifestations of the contradictions of capitalist socioeconomy. Resignification is directly connected with political semiotics, that is, the manifestation of power relations in different landscapes of meaning in which identities, practices and relations are constructed. One hegemonic interpretative order is predicated on the exclusion of other symbolic possibilities, while at the same time, counter-hegemonic practices of resignification can emerge and challenge the status quo. In the case of capitalist development, centred on commodification and production for exchange, socionatural relations are necessarily resignified and influence politico-economic dynamics (meanings are causes, according to the critical realist terminology).

In this sense, CPE can offer a good chance to connect the economic responses to culture with the resignification of the interrelated exploitation of society and nature. Resignification, away from the Western separation between science and ideology, is instrumental in removing the unhelpful cleavages between culture and nature (which also characterise Western culture). Likewise, moving beyond the micro-politics of post-structuralism—under the assumption that the specific and local experiences of minorities and marginalised groups retain the moral reservoir that has seemingly been lost in macro-politics and large-scale changes—resignification represents, first of all, a commitment to reinterpreting and helping to overcome a perverse reality that is systematically reinforced and cuts across nested scales of interaction. Furthermore, CPE can benefit from a 'conjunctural approach', along the lines proposed by Li (2014), which incorporates social, material and symbolic dimensions of any complex socioecological reality. According to this approach, a thorough examination must embrace practices, understandings, values and circuits within which capitalist relations of production and reproduction emerge. This essentially constitutes a neo-Marxist method of investigation that observes Marx's own epistemological position, in particular, the treatment of abstract concepts (e.g. labour, capital, land, etc.) not as ideal types but as categories that bring together concrete historical forms. As a result, culture does not imply a homogenising experience; rather, it is fundamental to consider the historical and geographical agency of individuals, social groups and also nature. By paying attention to 'conjunctures', the politics of resignification offers another powerful tool to renew and inform political economy's sensibility to culture (as intended with CPE).

Making use of the heuristic category of resignification, CPE thus has the ability to deal with the politics of nature—that is, supporting politico-ecological approaches—through an investigation into the double exploitation of society and of the rest of nature effected by the expansion of capitalism into new socio-spatial frontiers. This should be possible without the need to resort to the reductionism of post-structuralist perspectives. According to Peet and Watts (2004), in the final decades of the last century there was a postmodern obsession with reflexivity, which emphasised words more than deeds, discourse rather than development, hype over reality. Jones (2008) even argues that, due to the cultural turn and the influence of discursive revolution under post-structuralism, there was a common perception that this obsession was leading to a suspension of the political economy of nature (that is, political ecology). However, now that the post-structuralist moment has largely passed, academic research again needs to be grounded firmly in struggles over resources, which lie at the centre of struggles overpower. In this context, Peet and Watts (2004) advocate an agenda of critical studies and joint mobilisation defined as 'liberation ecologies'. We can even learn from the lessons of the New Left movement that emerged in the 1950s and 1960s, rejecting the 'economism' of both mainstream and left-wing parties, and advocating the incorporation of cultural questions, ecological challenges and grassroots politics instead of focusing only on the struggle to gain state power (Stevenson 2015).

This means that the full potential of CPE can be realised through a permanent and creative association between theory and practice, between critical philosophy and philosophising for change. CPE can be mobilised to help with the resignification of one of the most pressing issues, which is the meaning and basis of a revolution in the early twenty-first century. Bond (2016) and others describe revolution today as a transition to an eco-socialist society, where social, economic and ecological questions are considered together in the search for environmental justice. An eco-socialist transformation will require multiple forms of mobilisation, global alliances and ideological and practical prefigurations, which definitely provides space for the contribution of CPE. In that particular regard, Gramsci's elaboration on the philosophy of praxis, described particularly in Notebooks 10 and 11, is not only a provocative invitation to rupture and reconstruction, but his entire political project is significantly influenced by a cultural and historical awareness (Gramsci 1971). One particular chapter of that controversy is the fate of the Amazon and its catastrophic geography of modernisation, subjection and exploitation in the name of water development. It is, of course, impossible to deal with the full picture of water development in the Amazon here;

therefore, the rest of this chapter will briefly consider policies and strategies related to dam construction in the region. The intention is to make use of CPE, in particular, through the lens of resignification, to offer a commentary on the controversial politico-economic context of water in the Amazon, which has been firmly based on the simplification of socionatural systems according to the powerful discourse of resource exploitation, territorial occupation and economic growth led by the Brazilian State and its powerful economic allies. In schematic terms, it is possible to divide the history of dam construction and water infrastructure in the Brazilian Amazon into three main politico-economic periods (which have evident parallels with the phases of the agricultural frontier examined in previous chapters), as follows.

5.3 Dam Development in the Amazon: The Time of the Blunt Bulldozer

The above reflections on the limitations and prospects of CPE suggest that what is required is a more balanced, dialectical consideration of internal dimensions, that is, giving equal importance to the cultural, the political and the economic components of socioecological processes as they are lived and contested by different social groups. This would convert CPE into an even more heuristic approach that can be used to understand water development, as exemplified by the importance of the politics of resignification that underpins dam construction in the Amazon. It will become clear below that the disjuncture between the formal enunciation of water policies in the name of universal national development and the actual intricacies of their implementation produces a highly unequal distribution of opportunities and gains, which all depend on the affirmation of new meanings and values in the name of national modernisation. The formal and informal evolution of water management has been an integral element of the imposition of hegemonic ideologies and the mobilisation of labour and resources for the purpose of economic growth. As argued by Reis and Mollinga (2015), what is in place is a crucial dialogue between discursivity and materiality. The Amazon region, especially in Brazil, is the most recent frontier of resource grabbing (e.g. minerals, biodiversity, land, water, etc.) and increasingly a priority for hydropower development in South America. Many large new projects have been put forward and several have already been built, but all have suffered from a lack of transparency, participation, proper risk assessment and convincing economic analysis (revealing the contradictory side of dam construction, even according to mainstream economic reasoning); see, among others, Carvalho (2006), Fearnside (2016), Hanna et al. (2016), McCormick (2011), Scholz (2005).

These fundamental questions are not purely economic, political or cultural, but located precisely at the interface between those three realms. The problem, therefore, is one of cultural political economy par excellence. A discussion of the evolution of dam projects in the Amazon reveals a great deal about the contradictions of economic trends and the tragedy of the joint exploitation of society and the rest of nature covered by the hieroglyphics of capitalist modernity. As indicated by Marx, in this process concreteness and specificities are lost and the products of labour are converted into 'hieroglyphics' (in this case, hieroglyphics of modernity). Dams, as part of the infrastructure of development, are not only 'temples of the modern world' (paraphrasing Prime Minister Nehru), but are actually one of the most emblematic manifestations of certain politico-economic values at the expense of culture, ecology and resources for those affected by the advance of modernity. Over the centuries, many generations of Western explorers and naturalists have spent years, sometimes decades, collecting plants, animals and minerals, observing the locals and monitoring the river system, but the complexity and scale of the Amazon have left them unable to fully comprehend these. And, tragically, a large proportion of this richness is now being lost and disappearing fast, condemned to remain an enigma that never had the chance to be interpreted. The Amazon, along with much of the rest of South America, is too rich to be ignored, but too mysterious to be fully understood. In this epic and violent process, ignorance and greed prevail over science and justice. The Amazon region still defies interpretation but remains part of the overall 'project' of global modernisation that paves the way for the advance of capitalist relations of production and reproduction (Corrêa da Silva 2013). The category of 'nature' has assumed a particular meaning from the perspective of the modernising project, as it is considered to be a key element of native culture and, in that regard, operates to delegitimises the Western ('civilised') coloniser (Gidwani 2008).

Despite being the object of exploration and territorial disputes, for more than four centuries the economy of the Amazon was proto-capitalist in the form the mercantilist collection of 'hinterland plants' [*drogas do sertão*] during colonial years and, at the turn of the twentieth century, extraction of rubber from native trees. Land, minerals, biodiversity and water have always had great value for local populations—the rivers were the main means of communication, a source of staple food (fish) and the key reference for the opening of settlements and farms (mainly for subsistence agriculture)—but this translated into only modest monetary rewards. From the perspective of capitalist rationality (introduced early on by John Locke in relation to unused common lands shared by indigenous groups in the Americas), there was widespread wastage of resources (Gidwani 2008). Modernity, in the form of development, was considered the antidote to the supposedly primitive practices ('maladies') of traditional society and the waste caused by the mercantilist economy which was preventing the Amazon from fully integrating into the progress and industrial development of the south and southeast of Brazil. Contrasting with colonial and postcolonial times, when the treasures of the Amazon were captured and removed, state-led development required the consolidation of private property inside the region. Nonetheless, the same rentist strategy remained and instead of profit extracted from production, the process of development in the Amazon now relies on the appropriation of differential rent (in the Ricardian sense) from forested land and river systems (Ioris 2016). Development plans implemented in the Amazon since the 1950s have reproduced the old dreams of easy riches and disregard for local forms of culture and history beyond the immediate interests of accumulation (the classical pattern of pillage that defined America's conquest).

The first phase of state-led development was the time of the blunt bulldozer, between the 1960s and 1980s, when Brazil was ruled by a military dictatorship that basically worked to preserve conservative interests through a plan of centralised and autocratic modernisation. The incorporation of the Amazon region into technocratic economic plans was a deeply ideological process, formulated in the capital Brasília and encouraged by multilateral financial organisations. Instead of preserving the ecosystems and natural resources which are the fundamental basis for regional wealth, the type of development promoted constituted an attack on natural stability and on local populations. A major politico-economic vector associated with modernity was the construction of roads; but not just any roads—they had to be motorways of Amazonian proportions. The Transamazon highway, 5,400 km long, intended to ultimately connect the eastern and western sections of South America, was built through the forest and territory that had never been surveyed. The project took only four years and directly benefited from a period of fast economic growth (fuelled by foreign loans, state benevolence and the violent containment of political opposition). New farms began to open in 1966, particularly along the Belém-Brasília highway, the Transamazon highway, the BR-364 highway (Cuiabá–Porto Velho, which attracted more than 160,000 farmers every year during the 1980s) and the BR-163 highway (Cuiabá–Santarém, which is currently in the process of being paved, threatening to accelerate agricultural occupation even more). Settlers were encouraged to open farms in Amazonia not only through fiscal benefits but also by legislation that considered the removal of vegetation as 'improvement' of private property (Ioris 2016).

Agricultural expansion in Amazonia produced one of the greatest processes of land privatisation in the history of humanity. It was not only a social tragedy in terms of the loss of common resources and the proletarianisation of local populations, but also an ecological tragedy of planetary proportions. Large agricultural projects, some making use of millions of hectares, were aggressively promoted by the federal government, including through exploratory visits with leading businessmen and state ministers in 1973 (Branford and Glock 1985), which were not dissimilar to the one organised by the Stalin regime four decades earlier to the White Sea Canal. Agricultural colonisation projects along the new roads—where settlers were promised a plot of land and some financial and technological support (normally after great delays and broken promises)—represented an escape route for desperate farmers and labourers affected by recurrent droughts and structural water access inequalities in the semi-arid northeast of Brazil. Those coming from the Northeast Region often met landless groups evicted from the south of Brazil, where agriculture modernisation and latifundia prevented people from having access to land and resources. The quick, standard solution to poverty was to make the poor disappear by sending them to the Amazon. From an economic perspective, attracting destitute farmers to the region was not a successful policy, given that the settlers had no knowledge of the region, no means and no markets (Ioris 2017). Many returned to their areas of origin, moved to the cities or into slums along the roads, or found jobs in cattle ranches or plantation farms. We can already see here the perverse association between the lack of water, land and opportunities in the northeast, the lack of land and opportunities in the southern region, and the Amazon, with its abundance of water and land, but still no opportunities for the poor. The main issue was not the mere physical existence of water and land, but how it was mobilised as a resource according to social structures and politico-economic priorities.

Bulldozers were not only used to build roads, but also to divert rivers and erect dams for hydroelectricity generation. The quest for energy greatly helps to understand how and why the Amazon became increasingly important to national and international economic agendas, but without much regard for local and regional socioecological features. Particularly from the 1960s onwards, the energy sector in Brazil became highly centralised and controlled by an elite group of engineers and economists within the federal government, in cooperation with state administrations (Conca 2006). Most of the existing hydropower infrastructure in Brazil today was introduced by the military dictatorship, which benefited from the availability of international development money (petrodollars in particular) and the repression of social and political opposition. As a result of coordinated politico-ideological and economic strategies, around 74% (previously more than 90%) of electricity consumed in Brazil comes from hydropower schemes (ANEEL 2008), making the country one of the leading generators of renewable energy in the world (which can be considered a mixed blessing). The Amazon was not overlooked in the programme of dam construction. A region that makes up 54.4% of Brazilian territory and holds 78% of national freshwater was 'condemned' by the dictatorial government because of its geography to go through a process of water development (a risky endeavour considering the region's particularities, especially its extensive plains and complex socioecology). The Amazon, despite the region's massive amount of surface water (around a fifth of the planet's freshwater) is quite ill-suited to dams, and swaths of land and forest have been flooded in exchange for relatively little electricity.

The first projects were Coaracy Nunes in Amapá, and Curuá-Una in Pará, but the worst example of incompetence was Balbina dam, near Manaus, which flooded 2,360 km^2 for a very low power generation of only 112.5 MW and long periods of low operation due to low water flows (Fearnside 1989). As observed by the last author, the flat topography and small size of the drainage basin make output small, while vegetation has been left to decompose in the reservoir, resulting in acidic, anoxic water that will corrode the turbines. The project benefited from generous governmental subsidies and was carried out to supply electricity to Manaus, a large and fast growing capital city in the middle of the Amazon that far outpaced the contribution of Balbina. The Samuel dam also has a low efficiency, and the site of the dam is so flat that engineers have had to build 30 miles of dykes to help create a lake of 520 km^2; Samuel has the capacity to generate 217 MW, which was recognised even before its construction as insufficient to serve the growing cities of Porto Velho and Ji-Paraná. The national electricity utility Eletrobras conceded that Balbina and Samuel were mistakes and it was confirmed by one of the company's directors that 'air surveys and topography measurements were not accurately made because of the tree cover' (Simons 1989). The largest scheme built during this first phase was Tucuruí, which generates electricity particularly for aluminium smelting. Despite the significant level of social and environmental impacts, the decision-making in the case of Tucuruí was practically uninfluenced by environmental studies, which were done concurrently with the construction of the scheme (Fearnside 2001). During the filling of the Tucuruí reservoir, a large area of forest was not cleared and then died, leading to a large release of methane. With the construction of the Tucuruí dam, 2,430 km^2 of the forest was flooded and more than 33,000 people (besides the indigenous population) had to be resettled. The scheme also inundated part of three indigenous areas (Parakanã, Pucuruí and Montanha), the effect of which was added to the impact of transmission lines on this land; in addition, the artificial lake Tucuruí led to the extinction of various biological species and to the proliferation of waterborne illnesses (Comissão Pró-Índio de São Paulo 1991). Many of these problems remain unsolved and were present again during the next phase of dam construction.

5.4 The Phase of Politico-Economic Adjustment

The second phase of dam construction was the moment of adjustment, which coincided with the neoliberal reform of the Brazilian state and the rearrangement of national-developmentalism. The contradictions of the state-led model of development—implemented in Brazil during most of the twentieth century and intensified by the military governments making use of foreign savings through loans from multilateral banks—resulted in growing economic inefficiencies and reliance on the continuous injection of capital by the state. The military regime ended in 1985 in a context of political discontent, economic instability and great uncertainty. After a turbulent transition, and benefiting from the legitimacy earned since the presidential election of 1989 (the first in 29 years), the historical circumstances were ripe for pro-market reforms and the reorganisation of the state by the new government. But it was the macroeconomic stability and inflation control offered by the Real Plan in 1994 that provided the basis for legal and institutional adjustments. The Cardoso administration (1995–2002) promoted successful monetary and fiscal adjustments, which fuelled an ambitious reform of the state apparatus along the lines of liberalisation and growing integration into global markets. As in other comparable countries, a distinct neoliberal project was eagerly implemented (e.g. Spain and France) or maintained (e.g. the UK and Portugal) by centre-right political groups coalescing around the formerly leftist Cardoso.

The Real Plan was complemented by an extensive portfolio of institutional restructuring that included the removal of trade barriers (and at times the promotion of imports to avoid inflation), rigid monetary regulation, adjustments in the labour market and reorientation of the state apparatus. Energy policies and the electric sector were directly affected by the powerful neoliberalising agenda; systematic control of tariffs, widely used during the 1980s to contain inflation, was removed to favour private suppliers and a free energy market. Both generation and commercialisation of energy then became available to private national entrepreneurs, increasingly associated with international investors or energy companies. Water and environmental regulation were also transformed through the introduction of new legislation directly informed by the doctrine of ecological modernisation (such as the 1997 Water Law, centred on river basin committees, water licences and correspondent charges). In this sense, the water sector offers an emblematic demonstration of the choreography of continuities and changes that affected the economy and state regulation. In particular, the reorientation of government agencies and the introduction of new water legislation paved the way for the expansion of private gains extracted from publicly or collectively owned natural resources (Ioris 2009).

Because of the changing role of the state apparatus, which increasingly focused on regulation and policy-making rather than direct construction and operation (which nonetheless continued to happen, despite the neoliberal discourse), very few hydropower schemes were built during the second period. Coordination and decision-making became significantly diffused across many agencies, without the presence of a centralised, well-resourced agency as during the military dictatorship. This represented a real tension between ambitious efficiency and operational goals and the reality of institutional fragmentation and diminished investment by the national government (Goldenberg and Prado 2003). This tension is basically explained by the contrast between the immediate goals of private agents and the long-term demands of wider society. Also, the much-promoted virtues of the open energy market were never translated into investments and coordinated efforts. This culminated in the national energy crisis between June 2001 and March 2002 (by which time energy use was significantly reduced), caused by months of low rainfall, which required energy rationing, which came at a very high political cost (leading to the loss of the 2002 presidential election by Cardoso's political group). According to a National Energy Policy Council technical report (2001), the failure to invest in new dams was responsible for two-thirds of energy rationing (Kelman 2001). The energy and investment model introduced in the context of neoliberalising strategies was largely a failure, and even its advocates recognised that it had to be adjusted.

Nonetheless, as in other countries, there was a significant gap between the theory of the neoliberalising project and the concrete national experience. After nearly two decades of reforms, the Brazilian state remains fraught with ambiguities and internal conflicts, which ultimately reflect and incorporate the class-based antagonisms of civil society. The macroeconomic changes have produced winners and losers among the political elite, but to a large extent, the direction of Brazilian politics and the overall trends of development continue practically unaltered. The most vivid examples of continuities and path dependency are the public policies on poverty alleviation and environmental conservation. Despite compensatory measures and expanding environmental regulation, levels of inequality and ecological degradation have remained notably high, ultimately undermining the claims of beneficial economic growth and infrastructure expansion advanced by the national government and hegemonic groups. An examination of institutional reforms in the water sector reveals the complexity of innovation and continuity during the following decade.

5.5 Neo-Developmentalism, Neo-Liberalisation and Persistent Tensions

The third phase of this schematic analysis of dam construction in the Brazilian Amazon is the current period of neo-developmentalism and formalist environmental and socioeconomic regulation. This is the apex of the modernist project to date, as under complex environmental, labour and services legislation there have been sustained attempts to advance a new phase of increasingly large dams through associations between state agencies, construction companies, corporations and politicians. Between 2003 and 2016, under presidents, Lula and Dilma, national policies and government initiatives were characterised by a peculiar combination of populist measures (such as the conditional cash transfers of the *Bolsa Família* programme) with the reassurances offered to the financial and business sectors. During this third period, the country benefited from the commodity boom and increased its reliance on foreign goods, which came at the price of deindustrialisation and over appreciation of the national currency. In this context, the state revamped a selective programme of infrastructure construction in close alliance with engineering construction companies and local political leaders. National dam construction policies had to be adjusted because the best sites, situated closer to the main industrial and urban areas in the south and southeast of Brazil, were virtually all taken; consequently there were growing pressures to move into new areas, particularly in the Amazon, which are also more prone to controversies and impacts. Despite the mistakes of the past, the threat of hydroelectric exploitation of Amazonia has never been as present on the agenda as now, since the region allegedly holds around 50% of the national potential for electricity generation. Eletrobrás Plan 2010 lists 297 sites suitable for the installation of new plants in the country, of which 79 are located in Amazonia. The main areas for expansion are located on the Madeira River and waters flowing into the Tapajós River and on the Xingu and Tocantins rivers. In the Madeira Basin, after a lengthy political dispute, the Jirau and Santo Antônio plants were licenced in July 2007, allowing the overflow of up to 529 km^2. Project design and implementation were again highly controversial, facing clear opposition from the Bolivian government and environmentalists because of the superficial assessment of impacts, cosmetic implementation of the regulation and sustained influence by politicians and construction companies (Switkes 2008).

The contentious Belo Monte scheme, which has attracted great attention in the international media (e.g. 29 articles posted by BBC News alone between 2010 and 2017), demonstrates how politico-economic trends have persisted while the neoliberalising platform was being adjusted to fulfil neo-developmentalist goals. This represented the resumption of the construction of large hydroelectric power plants by the state apparatus and the encroachment of Brazilian energy demands upon neighbouring nations. Belo Monte was built along the Xingu River (famous for the large indigenous reservation in the upstream section) and is now the fourth largest power-station on the planet, just behind Itaipu (inaugurated in 1984 on the border between Brazil and Paraguay) and Xiluodu and Three Gorges in China. The Belo Monte project, under another name, was originally conceived by the military in 1975 (aiming to flood, considering the six planned dams along the Xingu River, a total of 14,500 km^2) but had no chance to go ahead due to the regime's growing financial problems and ultimate collapse. The residents of Altamira (the site of the dam) and local indigenous groups have maintained an organised resistance (although condemnation was not uniform, but showed some disagreement between opposing groups) influenced by the traumatic experience of Tucuruí. There was a large gathering in 1989 with more than 1,000 participants, including more than 600 indigenous people, which attracted international attention to the dispute and led to the cancellation of a World Bank loan under negotiation (Carvalho 2006). As a result of the protest, the initial design was changed: the area designated for flooding decreased to 400 km^2 and the name of the dam was changed to Belo Monte.

The project was again modified by the Lula government, elected in 2002, which ironically had as top energy authorities many of those who had opposed this and other similar projects in the past. Capacity and transmission lines were reduced, and the new project removed the large reservoir in order to minimise negative impacts. However, the controversial features of Belo Monte continue to stir protests and serious resistance. The granting of environmental and water licences in 2011, as well as public consultation required for the approval of the project, was notoriously undemocratic and aggressively pushed forward by the federal government under the justification that economic growth required additional sources of energy (Sevá Filho 2005). Abers et al. (2017) argue that the government's tendency to prioritise the interests of the construction and business sectors in new infrastructure projects in the Amazon, together with its persistent disregard for public participation and social demands, reveal the contradictory and asymmetric priorities of the state. A consortium of state-owned companies called Norte Energia won the contract to build Belo Monte and manage it for 35 years. Despite the rationalisation of the engineering design, it would still displace peasants and indigenous tribes, affect river ecology and the water regime. The project was the object of a lengthy battle in the courts, which led to repeated interruptions of its construction and operation (the last interruption ordered by a judge occurred in 2017). Notwithstanding the political struggle, the dam was inaugurated in 2016, with 11,233 MW of installed capacity and at a cost of more than US$ 13 billion. Belo Monte prompted a negotiation with Chinese investors to instal an aluminium factory in the region of Barcarena (in the state of Pará) and with the Canadian company Belo Sun to extract gold in the area around the hydropower plant (the environmental licence was granted in March 2017, but critics such as Amazon Watch say that the risks were ignored by the authorities; Poirier 2017), among other similar projects. The most controversial issue relates to the actual viability of the project without other supporting dams upstream. Due to the long dry season and the resulting long period of low flows in the Xingu River Basin (a common feature of eastern Amazonian rivers), Belo Monte has a low operation and economic performance if operating as a single dam. This means that a cascade of dams is necessary, but this would multiply the impacts on indigenous land, natural parks and farms.

The national government has guaranteed on many occasions that Belo Monte will be the only large-scale dam on the Xingu River, but the wider problem for Brazilian society is the low level of government legitimacy and eroded trust in public authorities. This is a global phenomenon that seriously impacts the quality of formal democracy in many countries, Brazil included. Similar reassurances were given in relation to the construction of other schemes in the Amazon, such as the 43 large dams to be implemented, even more controversially, along the even more vulnerable Tapajós River by 2022 (Fearnside 2015). Those dams are directly benefiting from the lenient interpretation of the legislation and lax enforcement of water and environmental regulations, and the result is that several indigenous reserves and conservation areas will be flooded and degraded. The insistence on the construction of those dams reveals the incompatibilities between public policies (stemming from an authoritarian model of governance by the electrical sector, construction companies and private industries), but also the creative and persistent resistance of indigenous peoples and other groups advocating for defence of their territories, livelihoods and culture (Alarcon et al. 2016). For many activists and local communities, there is a distinct feeling of betrayal and deception, especially because of the surprising new alliances formed between elected left-wing politicians and the traditional economic players and conservative politicians (Melo 2016). After more than 500 hundred years of the pillage of indigenous lands and destruction of ecosystems, a new hope emerged in Brazil as Lula, a union leader, was elected as president. However, despite great expectations, he then allied his government (2003–2010) with the most archaic and corrupt politicians and construction companies in the country. During the Dilma government (2010–2016), the secret service was asked to spy on indigenous leaders, anthropologists and social movements who opposed the construction of Belo Monte (Valente and Aragon 2017). 'The betrayal is hard to understand. But time doesn't care if you understand' (Diversi 2014: 243). Even more disturbing were the dishonest practices employed during the construction of Belo Monte and in many other projects championed by Lula and his loyal successor Dilma (who was impeached and removed from office in 2016), including oil and nuclear developments, football stadiums and the Olympic Games.

Corruption and manipulation on an industrial scale is only now being revealed through the work of the federal police and the public prosecutor in the *Lava-Jato* [Carwash] Operation. Many politicians, operators and directors of engineering companies have decided to collaborate with the judicial authorities in order to secure reduced criminal penalties (plea bargain agreements), and a monumental scheme has been exposed involving bribery, illegal financing of electoral campaigns, flawed costing and engineering details, irregular granting of environmental and water licences, and cartelisation of the construction sector (Calixto 2017). With increasing numbers of arrests among top politicians, senior civil servants and owners of construction companies, including a significant number of members of congress, such as the speakers of the Senate and of the House, and several ministers (Casado et al. 2017), it has become clear that corruption in the water and energy infrastructure in the Amazon is deeply rooted in the multiple operations conducted by the national state in the name of development, but that they mainly favour the interests of established economic and political groups based in the core economic areas (far away from the socio-spatial frontiers in the Amazon). A similar approach, involving the same construction companies (most notoriously Odebrecht, South America's largest construction group), was responsible for the construction of hydropower projects in other countries, as in the case of the Lauca dam in Angola (Leahy 2016). Corruption is a global problem that feeds on national and local opportunities, as in the case of the new rounds of hydropower construction and mineral extraction in the Amazon.

The construction of new dams has many ramifications, which multiply the repercussions of corruption beyond the mere appropriation of public funds. The most significant is the connection between dams and agribusiness. One key example of this is that the new dams will allow large-scale barges to pass rapids and cross waterfalls, which will significantly reduce transport costs to ports along the Amazon River, making them accessible to transcontinental transportation ships, and ultimately increasing the profitability of soybean cultivation and the areas available for such production (Fearnside 2015). Justification for the new hydropower schemes seems to be less about energy generation and more about the creation of new navigation routes. At face value, it seems that there is great potential in the expansion of river navigation, considering that this currently only accounts for 4% of domestic Brazilian trade, but it is almost certain that navigation plans will follow speculation and financialisation pressures, leading to land grabbing, chaotic migration and deforestation (Becker 2012). The growing politico-economic importance of agribusiness in Brazil is directly connected with the challenges of water management and Amazon development. The sector is commonly considered a great Brazilian achievement, due to technological improvements and production growth, but it is also responsible for ecosystem degradation, contamination of water reserves, socio-spatial inequalities and macroeconomic vulnerability (Ioris 2015). An apparent techno-economic success has encouraged further land and water grabbing, as the agribusiness sector has been unable to resolve long-lasting problems such as economic development that benefit small groups of large-scale landowners, commercial companies and transnational corporations, with wider society paying the price in the form of mounting environmental and social impacts.

5.6 Conclusions

The large-scale and rapid incorporation of water into economic development strategies, as examined above, demonstrates the persistent and systematic advance of modernity in the Amazon. It has been a phenomenon of epic proportions, involving both material and structural transformations of ecosystems, communities, lifestyles and socio-economic arrangements. At the centre of those processes of change lies the appropriation of territorial resources and the conversion of the complexity and heterogeneity of the Amazon into the abstracted realm of market transactions and the circulation of capital. It has involved and been based on the privatisation of common resources, shared spaces and opportunities. Essentially, it has been the tragedy of the non-commons and the widespread commonality of tragedy. One of the disturbing facets of this broader trend is the instrumentalisation of water, beyond its socio-cultural, economic and geomorphological functions, to fulfil exogenous demands for energy, navigation and agricultural expansion. Development largely depends on water, but it must first be recreated as a resource divested of Amazonian features.

In that regard, CPE can be another helpful intellectual tool in the examination of the controversial and contradictory basis of regional development and water management in areas of rapidly advancing socio-spatial frontiers. Among many positive attributes, CPE can help us to understand the impact of hegemonic pro-development ideologies and discourses, as well as to comprehend the complexity of the interface between economic and more than economic practices. The main conclusion here is that the examination of problems and the search for alternatives requires the proper resignification of ongoing economic trends and associated policies. Without patronising or romanticising traditional populations, it is important to recognise that these groups are particularly well equipped to understand the vital importance of seasonal hydrology for their survival and social organisation. The resignification and revaluation of their knowledge and techniques is an integral part of the critique of hegemonic water development. As inheritors of the knowledge accumulated by their ancestors, the populations dwelling along the banks of rivers possess a profound affinity with these rivers, the soils, and the biodiversity of Amazonia. Through an intense interaction between society and nature, traditional populations have learned to respect natural cycles and preserve the ecology of the rivers. Nevertheless, despite the fact that these populations have historically lived sustainably with regional nature, the technocrats of development tend to regard them as largely marginalised sectors of society who may have something to gain in an uncertain future if the current mechanisms for exploiting Amazonia are increased even more. The centralised and arrogant perspective of official bureaucrats and businesspeople prevents them from taking the real circumstances of these populations into account with regard to public policies and water management in particular. While it will not be officially recognised, the misfortune of these populations overwhelmingly affected by the process of economic growth will continue to be a symbolic test of the deep contradictions of modernity and development in the Amazon.

References

Abers RN, Oliveira MS, Pereira AK (2017) Inclusive development and the asymmetric state: big projects and local communities in the Brazilian Amazon. J Dev Stud 53(6):857–872

Alarcon DF, Millikan B, Torres M (eds) (2016) Ocekadi: Hidrelétricas, Conflitos Socioambientais e Resistência na Bacia do Tapajós. International Rivers Brasil, Brasília/Universidade Federal do Oeste do Pará, Santarém

Althusser L (2006) Philosophy of the encounter: later writings, 1978–1987 (trans: Goshgarian GM). Verso, London and New York

ANEEL (2008) Atlas de Energia Elétrica do Brasil, 3rd edn. Brasília, Agência Nacional de Energia Elétrica

Badiou A (2012) Philosophy for militants (trans: Bosteels B). Verso, London and New York
Becker BK (2012) Reflexões sobre Hidrelétricas na Amazônia: Água, Energia e Desenvolvimento. Boletim do Museu Paraense Emílio Goeldi (Ciências Humanas) 7(3):783–790
Bond P (2016) South Africa's next revolt: eco-socialist opportunities. In: Panitch L, Albo G (eds) Rethinking revolution—socialist register 2017. Merlin Press, London, pp 161–185
Bourdieu P (2004) Science of science and reflexivity (trans: Nice R). Polity, Cambridge
Branford S, Glock O (1985) The last Frontier. Zed Books, London
Calixto B (2017) O Que as Delações da Odebrecht dizem sobre Corrupção nas Hidrelétricas da Amazônia. Revista Época, 13 Apr 2017. http://epoca.globo.com/ciencia-e-meio-ambiente/blog-do-planeta/noticia/2017/04/o-que-delacoes-da-odebrecht-dizem-sobre-corrupcao-nas-hidreletricas-da-amazonia.html
Carvalho GO (2006) Environmental resistance and the politics of energy development in the Brazilian Amazon. J Environ Dev 15(3):245–268
Casado L, Mattoso C, Megale B (2017) Ao TSE, Delator cita Lobão e PMDB em Esquema de Propina de Belo Monte. Folha de São Paulo, 21 Feb 2017. http://www1.folha.uol.com.br/poder/2017/03/1869389-ao-tse-delator-cita-lobao-e-pmdb-em-esquema-de-propina-de-belo-monte.shtml
Clarke S (1991) Marx, marginalism and modern sociology: from Adam Smith to Max Weber, 2nd edn. Palgrave Macmillan, London
Comissão Pró-Índio de São Paulo (1991) Segundo Tribunal Internacional das Águas. Caso: As Hidrelétricas na Amazônia Brasileira. São Paulo
Conca K (2006) Governing water: contentions transnational politics and global institution building. MIT Press, Cambridge and London
Corrêa da Silva M (2013) Metamorfoses da Amazônia, 2nd edn. Manaus, Valer
Diversi M (2014) Damming the Amazon: the postcolonial march to the wicked West. Cult Stud Crit Method 14(3):242–246
Eagleton T (2011) Why Marx was right. Yale University Press, New Heaven and London
Fearnside PM (1989) Brazil's Balbina dam: environment versus the legacy of the pharaohs in Amazonia. Environ Manag 13(4):401–423
Fearnside PM (2001) Environmental impacts of Brazil's Tucuruí dam: unlearned lessons for hydroelectric development in Amazonia. Environ Manag 27(3):377–396
Fearnside PM (2015) Amazon dams and waterways: Brazil's Tapajós Basin plans. Ambio 44:426–439
Fearnside PM (2016) Environmental and social impacts of hydroelectric dams in Brazilian Amazonia: implications for the aluminium industry. World Dev 77:48–65
Figes O (2008) The whisperers: private life in Stalin's Russia. Penguin, London
Fraser N, Honneth A (2003) Redistribution or recognition? A political-philosophical exchange (trans: Golb J, Ingram J, Wilke C). Verso, London and New York
Gidwani V (2008) Capital, interrupted. University of Minnesota Press, Minneapolis and London
Goldenberg J, Prado LTS (2003) Reforma e Crise do Setor Elétrico no Período FHC. Tempo Soc USP 15(2):219–235
Gramsci A (1971) Selections from the prison notebooks (trans: Hoare Q, Smith GN). Laurence and Wishart, London
Hanna P, Vanclay F, Langdon EJ, Arts J (2016) The importance of cultural aspects in impact assessment and project development: reflections from a case study of a hydroelectric dam in Brazil. Impact Assess Proj Apprais 34(4):306–318
Hegel GWF (2008[1821]) Outlines of the philosophy of right (trans: Knox TM). Oxford University Press, Oxford
Ioris AAR (2009) Water reforms in Brazil: opportunities and constraints. J Environ Planning Manag 52(6):813–832
Ioris AAR (2010) The political nexus between water and economics in Brazil: a critique of recent policy reforms. Rev Radic Polit Econ 42(2):231–250
Ioris AAR (2015) The production of poverty and the poverty of production in the Amazon: reflections from those at the sharp end of development. Capital Nat Soc 26(4):176–192

Ioris AAR (2016) Rent of agribusiness in the Amazon: a case study from Mato Grosso. Land Use Policy 59:456–466
Ioris AAR (2017) Encroachment and entrenchment of agro-neoliberalism in the Centre-West of Brazil. J Rural Stud 51:15–27
Ioris AAR (2018) Assessing freshwater sustainability at the river Basin scale. Scholars' Press, Riga
Jeannot T (2010) The enduring significance of the thought of Karl Marx. Int J Soc Econ 37(3):214–238
Jessop B, Sum N-L (2006) Beyond the regulation approach: putting capitalist economies in their place. Edward Elgar, Cheltenham
Jessop B, Oosterlynck S (2008) Cultural political economy: on making the cultural turn without falling into soft economic sociology. Geoforum 39(3):1155–1169
Jones S (2008) Political ecology and land degradation: how does the land lie 21 years after Blaikie and brookfield's land degradation and society? Geogr Comp 2(3):671–694
Kelman J (ed) (2001) Relatório da Comissão de Análise do Sistema Hidrotérmico de Energia Elétrica. Brasília, Conselho Nacional de Política Energética
Kirsch S (2012) Cultural geography I: materialist turns. Prog Hum Geogr 37(3):433–441
Leahy J (2016) A Brazilian bribery machine. Financial Times, 28 Dec 2016. https://www.ft.com/content/8edf5b2c-c868-11e6-9043-7e34c07b46ef
Li TM (2014) Land's end: capitalist relations on an indigenous frontier. Duke University Press, Durham and London
Mann G (2012) Release the hounds! The marvelous case of political economy. In: Barnes TJ, Peck J, Sheppard E (eds) The Wiley-Blackwell companion to economic geography. Blackwell, Oxford, pp 61–73
Marx K (1975) Early writings (trans: Livingstone R, Benton G). Penguin and New Left Review, London
Marx K (1976[1867]) Capital: a critique of political economy, vol 1 (trans: Fowkes B). Penguin, London
Marx K (1988[1930]) Economic and philosophic manuscripts of 1844 (trans: Milligan M). Prometheus Books, Amherst, NY
Marx K, Engels F (1974[1845–1846]). The German ideology (trans: Arthur CJ). Lawrence and Wishart, London
McCormick S (2011) Damming the Amazon: local movements and transnational struggles over water. Soc Nat Resour 24(1):34–48
Melo MA (2016) Crisis and integrity in Brazil. J Democr 27(2):50–65
Ollman B (2003) Dance of the dialectic: steps in Marx's method. University of Illinois Press, Chicago
Peet R, Watts M (eds) (2004) Liberation ecologies: environment, development, social movements, 2nd edn. Routledge, Abingdon
Poirier C (2017) Toxic mega-mine looms over Belo Monte's affected communities. http://amazonwatch.org/news/2017/0404-toxic-mega-mine-looms-over-belo-montes-affected-communities
Reis N, Mollinga PP (2015) Public policy and the idea of the vietnamese state: the cultural political economy of domestic water supply. Asian Stud Rev 39(4):628–648
Sayer A (1992) Method in social science: a realist approach, 2nd edn. Routledge, London and New York
Sayer A (2001) For a critical cultural political economy. Antipode 33(4):687–708
Scholz I (2005) Environmental policy cooperation among organised civil society, national public actors and international actors in the Brazilian Amazon. Eur J Dev Res 17(4):681–705
Scott JC (1998) Seeking like a state. Yale University Press, New Heaven and London
Sevá Filho AO (ed) (2005) Tenotã-Mõ: Alertas sobre as Conseqüências dos Projetos Hidrelétricos no Rio Xingu. International Rivers Network, São Paulo

Simons M (1989) The Amazon forest: Brazil wants its dams, but at what cost? The New York Times, 12 Mar 1989. http://www.nytimes.com/1989/03/12/weekinreview/the-world-the-amazon-forest-brazil-wants-its-dams-but-at-what-cost.html

Stevenson N (2015) Post-citizenship, the new left and the democratic commons. Citizsh Stud 19(6–7):591–604

Sum N-L, Jessop B (2013) Towards a cultural political economy: putting culture in its place in political economy. Edward Elgar, Cheltenham

Switkes G (ed) (2008) Muddy waters: consequences of daming the Amazon's principal tributary. Berkeley, CA, International Rivers

Valente R, Aragon R (2017) Abin Espionou Indígenas e ONGs no Governo Dilma. Folha de São Paulo, 9 May 2017. http://www1.folha.uol.com.br/poder/2017/05/1882257-abin-espionou-indigenas-e-ongs-no-governo-dilma.shtml

Chapter 6
Production of Poverty and the Poverty of Production in the Amazon

Abstract It is discussed how socio-natural interactions in the Amazon bring the imprint of old and new forms of injustice, which are central driving forces in the reshaping of landscapes according to the balance of political power. Poverty is a condition of unsatisfied material and sociopolitical needs caused by combined mechanisms of exploitation, alienation and exclusion. The prevailing model of development and environmental management systematically reinforces hardship and destitution, at the same time it allows the corrosion of the forest. The chapter focuses on the lived experience of extractivist communities in the eastern Amazon, approaching these from the perspective of groups who only marginally benefit from the process of frontier making. Empirical results show that the politics of development and poverty in the Amazon do not happen about or around the forest, but with and through the forest. The main conclusion is that development is based on anticommons strategies, poverty is the outcome of the exercise of political hegemonies exercised over socio-nature, the systematic failure of poverty alleviation approaches and that forest ecosystems are also central players in the whole process of social differentiation and political resistance.

Keywords Para · Tapajos · Extractivism · Gender · Socio-nature

6.1 The Poverty of Development in the Amazon Frontiers

As we have seen in the previous chapters, a few regions in the world have given rise to so much politico-ecological controversy and been associated with such high levels of uncertainty like the Amazon. Since the time of Francisco de Orellana (ca. 1511–1546), and his epic search for El Dorado, the Amazon has been known for an 'extravagant' geography, immense challenges and, potentially, even greater rewards. The region was considered the archetypical representation of the Garden of Eden by renaissance chroniclers and generations of explorers (Holanda 2000). Yet, after the economic boom because of a highly profitable rubber production at the turn of the twentieth century, there was an inescapable reversal to subsistence agriculture and barter economy (Bunker 1985; Silva 2012). A few decades later, in the post-World War II period, the Amazon became one of the most disputed frontiers of

A. A. R. Ioris, *Frontier Making in the Amazon*, Key Challenges in Geography,
https://doi.org/10.1007/978-3-030-38524-8_6

Western modernity, a process that not only engulfed but also recreated, territories, relationships and peoples. Particularly in its Brazilian section, new development-related initiatives were put into effect by the military dictatorship which resulted in an increasing conversion of catchments and localities into hotspots of intense commodity production (Pinto 1980). The promise of rapid enrichment, often combined with cultural estrangement and sheer fascination, provided once again the rationale for violent conquest, eviction of existing communities and the expropriation of land, resources and livelihoods (Gondin 2007). Hegemonic relations of production and reproduction have deliberately disregarded ecological limits and aggressively incorporate nature into the logic of commodity production exactly because of the money to be made from the privatisation of collective ecosystems and territorial resources (Ioris 2007). Far from being politically neutral, the product of such changes has been primarily accumulated in the hands of a coalition between traditional elites and emerging business sectors that are endorsed by local and national public authorities.

The aim of this chapter is to expand our understanding of the meaning and consequences of a poverty making economy at the frontiers of the expanding pressures of national and regional development. Such critical discussion is important because, as pointed out by Santos (2014: viii), there is no global justice without global cognitive justice (as much as 'there is no way of knowing the world better than by anticipating a better world'.). Instead of the mere absence of material means, empirical results will show that poverty is a relational phenomenon that arises from the selectiveness of productive activities and socio-economic opportunities. Persistent levels of poverty and new structural inequalities are nothing else but the mirror image of development based on short-term gains, lasting negative impacts and commodification of nature. More importantly, poverty constitutes an integral feature of socioecological (or socio-natural) interactions (after Castree 2002) which are deeply politicised and encapsulate class-based differences and the balance of power. The interactions between society and (the rest of) socio-nature bring the imprint of old and new forms of injustice, which are central driving forces in the constant reshaping, and contestation, of space. For instance, governmental instruments (such as credit, subsidies and the granting of private property) and infrastructure investments (in the form of roads, ports and warehouses) attracted different contingents of people to the Amazon, who have only marginally benefited from the process of development. At the same time, Amazonian biodiversity has been filled with inequalities and asymmetries spreading from local to regional and even international scales of socioecological exchange. The reality is that capital, as a dominant social relation (Marx 1976), encroaches upon the Amazon to retransform the landscape, generate serious social and environmental impacts, and create new signifiers according to its own priorities. It does not matter if the forested land is crucial for the survival of people and ecosystems, because land [i.e. nature], as capital can be exploited and increased 'just as much as all the other instruments of production'. (…) The tragedy derives from the fact that '*[l]and as capital is no more eternal than any other capital*' (Marx 1956: 185; emphasis added).

In the text, the basis and the genesis of poverty in the lower Tapajós River Basin (a tributary of the Amazon River with headwaters in the state of Mato Grosso, where soybean production and hydropower schemes continue to increase, as discussed in other chapters), which is one of the Brazilian areas under particularly intense development pressures (e.g. plantation farms, hydropower, mining, roads and navigation) will be specifically revisited. The discussion presented in this chapter is based on fieldwork carried out between 2010 and 2011, with additional follow-up visits in subsequent years, undertook to investigate the mechanisms of power distribution and the environmental (i.e. socioecological) contradictions of mainstream policymaking. The starting point was the recognition that poverty is a location-specific phenomenon that directly reflects foreign influences and the intervention of the national government (Pogge 2008). In addition, the research required the systematic study of the interconnections between environmental change and the everyday experiences of hardship in the areas under study. The main unit of analysis was precisely the relationships between people, things and processes, that is, the reality should be conceived relationally (these relations cover and include what is related) and the condition for the existence of things and society is taken to be part of what they are (Ollman 1976). The research had also to consider different materialities and symbolic constructions of deforestation in areas which have specific issues related to poverty and development. Such methodological approach facilitated the investigation of the meaning of poverty and human vulnerabilities in the context of environmental changes and ecosystem degradation. Research findings will demonstrate that the politics of development and poverty in the Amazon do not happen about or around the forest, but with and through the forest. It will make evident that development is not a monolithic phenomenon, but it is full of cracks and intricacies where political reactions can flourish.

One of the main goals of the project was to give voice to 'forest-dependent people' (i.e. low-income communities living in close contact with Amazon ecosystems and who marginally benefit from the process of development) in order to articulate their concerns over degradation and deforestation. The intention was to go beyond the usual stereotypes that typically portray the poor as passive, disengaged from government plans or guilty of forest degradation and social unrest. The research methods included participant observation and engagement with rural communities, one regional workshop (attended by local government officials, community development officers, community representatives, research institutions and NGOs), analysis of documents and policies, and almost one hundred semi-structured interviews (respondents agreed to the disclosure of their names). It basically followed a participatory action research approach oriented towards social change (Kindon et al. 2010) through a collaborative interaction especially with residents of the extractive reserve (RESEX), Tapajós-Arapiuns near Santarém. The reserve, on the left margin of the lower Tapajós River, was established in 1998 and contains 72 communities (with around 18,000 people, the majority are descendants of migrants from the drought-ridden Northeast of Brazil who came to the Amazon to work in the rubber trade and eventually settled in the region); their reliance on extractive activities, the production of artisanal goods and subsistence farming provide an evocative contrast to the growing number of soybean farms, miners and timber companies in the region. Through a persistent dialogue with people living in and near the forest in the case study areas (Fig. 6.1) it became clear that local concepts of poverty are, for many, associated with urbanisation, restricted access to cultural opportunities, lack of political space, and

Fig. 6.1 Meeting with residents of the extractive reserve Tapajós-Arapiuns

destruction of ecosystems and, ultimately, the Amazon at large. Before examining the opinions and experiences of those at the sharp end of development, it will be necessary to briefly describe the complex geography of the Tapajós.

6.1.1 Developing and Impoverishing the Tapajós River Basin

This section will examine the main drivers of regional development and the most relevant socioecological pressures in the Tapajós River Basin, an area with around 490,000 km^2, relatively well-preserved forests and home to only one million people. Figure 6.2 shows the location of the Tapajós in the eastern section of the Amazon (in the states of Pará and Mato Grosso)[1] and its two forming rivers, Juruena and Teles Pires. The influence of global market demands is certainly now new, as the region played an important role in the history of colonisation and natural resource exports. The Portuguese occupation of the Tapajós basin started in the 1630s and the main settlement, Santarém, was founded in 1755. Later, at the turn of the twentieth century, it was the main producer of natural (vegetal) rubber needed primarily by the car industry. Rubber was the first agro-industrial business in the Amazon and responded for the circulation of significant sums of capital, which were nonetheless accumulated in the hands of the powerful tradesmen in Manaus and Belém (demonstrated by the

[1]A small tract of land is in the State of Amazonas in the municipality of Maués.

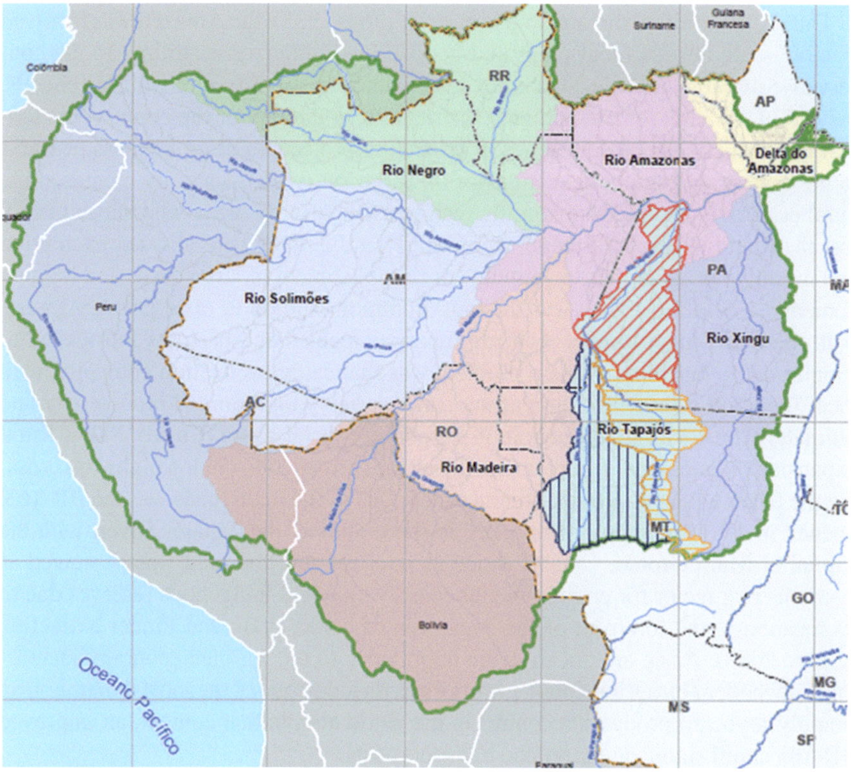

Fig. 6.2 Location of the Tapajós in the Amazon in Mato Grosso and Pará and its two upstream sub-basis (River Juruena [vertical stripes] and River Teles Pires [horizontal stripes]) (image adapted from Grupo de Estudos Tapajós)

exuberant architecture of the period), international banks and corporations and even geopolitical disputes between the various producing countries, as in the case of Peru and Brazil (Hecht 2013). That led Henry Ford to embark upon an ambitious project of high-tech rubber tree cultivation, which in a few decades became a monumental failure and was eventually closed down in 1945 with great economic losses (Dean 1987). Despite Ford's unrivalled capitalist brilliance, the enterprise carried out in the Tapajós River Basin was notoriously inadequate due to the stubborn disdain for local agroecological and sociocultural features. With the virtual collapse of commercial rubber production, those involved in the extraction and commercialisation either left the region or turned back to subsistence activity and had to rely on forested ecosystems for their survival. What is relevant for the present discussion is to note that their condition did not necessarily get worse, but the locals were able to subsist exactly because of their socioecological knowledge and multiple interconnections. Living in close contact with the forest did not constitute a position of hopeless vulnerability and poverty, but represented a concrete alternative in a context of structural inequalities and sustained socioecological exploitation (i.e. of nature and the workforce).

During the military dictatorship in Brazil (1964–1985), the Amazon was involved in large-scale development plans and geopolitical initiatives according to the conservative ideology and the conspiracy fears (almost paranoid) of the ruling generals. Instead of agrarian reform and land distribution in the rest of the country, the federal government advanced a series of relocation projects to bring landless peasants to the eastern and southern sections of the Amazon (who essentially increased the number of poor social groups in the region, especially by those who subsequently lost their land to larger farmers). More importantly, there were also tax exemption and incentive programmes to stimulate the establishment of large farms and logging companies. Slash and burn was the most common method to open primary forests with the high-valued timber sold to timber companies and pastures introduced to support cattle ranching. A main problem was that cattle production competed with local food and subsistence agriculture and invariably undermined biodiversity and ultimately the social reproduction of the poorest social strata (Castro 2007). Rural expansion policies were complemented with the construction of long highways crossing the forest and integrating different states. One of the main roads was the BR-163, opened in 1974 to connect Santarém, in the mouth of the Tapajós River, with the capital of Mato Grosso, Cuiabá. Since its construction, the road was perceived as a serious risk factor for environmental conservation and indigenous reserves due to its location at the forefront of the expansion of cattle, grain and timber extraction (Castro 2008). As an area of strategic importance to the national economy increasingly dependent on agribusiness, Mato Grosso is today one of the most dynamic crop (mainly soybean) production centres in the world and, in that context, an improved BR-163 could significantly reduce transport costs.

With the increase of land prices in Mato Grosso and other Brazilian states, the lower Tapajós offered a new window of opportunity for the same phenomenon to take place once again. More than 60 years after the end of rubber plantation by Henry Ford, soybean began to be cultivated around Santarém, as the result of a convergence of cheap land, government support, technological improvements and flexibilisation of environmental regulation (Pereira and Leite 2011). Soybean, as one of the main commodities exported by Brazil today, is highly emblematic of the type of development that is being promoted and, crucially, the politico-economic priorities and the hierarchy of benefits and beneficiaries (Ioris 2018; Freitas et al. 2017; Rivero and Jayme Jr. 2008). The cultivation of soybean in the lower Tapajós has increased substantially after the soy giant Cargill constructed a controversial export harbour terminal in Santarém in 2003, which was associated with conflicts and court disputes since its announcement and turbulent approval (Costa 2012). See Fig. 6.3. In 2014 alone, more than two million tonnes of grains were exported through the port of the corporation in Santarém. Upstream along the Tapajós River, in a location known as Miritituba, the TNC Bunge started to operate other navigation facilities to ship soybean to the international ports located at the mouth of the Amazon River.[2] It

[2]In 2014, more than ten other companies were avidly waiting to be licenced to start naval operation in Miritituba.

Fig. 6.3 Cargill Soybean terminal in Santarém

is expected that the so-called 'north exit' for soybean produced in the Amazon—combining better roads and fluvial navigation (one of the main terminals, Miritituba, has the capacity to export 16.5 million tonnes every year; in 2017, only 7 million tonnes were shipped from this port because of logistical bottlenecks and the difficult access by road)—could reduce transport costs by 30% (when compared to ports in the southeastern coast of Brazil). Government plans for the expansion of river navigation, connecting with it new railways and better roads, are very ambitious and have attracted the interest of foreign investors and operators, particularly Chinese. Yet, there are crucial parallels between rubber and soybean farming, in the sense that are both activities that rely on the disorganisation of existing lifestyles, on the predatory appropriation of nature and even on racist, discriminatory attitudes of newcomers against the locals, non-white populations (Costa 2012).

Farming expansion followed a pattern of (legal and illegal) occupation of cheap land, production intensification and then migration to new areas. The purpose of land grabbers (commonly identified as *grileiros*) is to secure massive gains from speculation and sudden increases in the price of land. Farmland around Santarém had the highest percentage of price increase in 2010: between 88 and 111% in one single year (FNP 2011, quoted in Costa 2012). Land grabbing in Brazil is unique as it involves both foreign and national capitals mobilised through the state ideology of development and national security (Oliveira 2013). Roads, government incentives and weak rule of law have provided fertile ground for the perpetuation of land grabbing, which has been a widespread practice in the Tapajós by faking titles, corrupting officials and unrestricted use of violence. Only between 1996 and 2001, 475 social activists were murdered in the state of Pará (London and Kelly 2007). For instance, in 2005, the American Catholic Sister Dorothy Mae Stang, a protector of the local poor and the forest, was murdered after repeated threats from loggers and landowners. The escalation of land price only aggravates the displacement and violence against peasants and indigenous groups. Violent conflicts are common between gold miners and indigenous groups, as in the case of the Mundurukus who still wait for the regularisation of their land. On 28 November 2014, a delegation occupied the office of the Brazilian indigenous service FUNAI in Itaituba to complain about the situation and, with no solution in sight, decided to initiate the demarcation themselves, obviously leading to more controversy.

There is a widespread sense of lawlessness and persistent tensions associated with land tenure disputes, illegal deforestation and the exploitation of forest and mineral resources (Zaitch et al. 2014). The Tapajós is also the most important gold panning area in Brazil. Before 1978, the activity was carried out as rudimentary mining, but production then increased with the introduction of machines and motors. The increasing price of gold in recent years only attracts more artisanal gold miners [*garimpeiros*] to the region. In addition, there is growing pressure for the construction of new hydroelectricity schemes in the Amazon, and the Tapajós in particular, which are considered the main untapped reserve of renewable energy in Brazil. The contentious, inefficient and hugely expensive Belo Monte dam is currently under construction in the Xingu River (a project constantly stopped due to lawsuits, legal disputes and contract renegotiations), but initial assessments indicate that the Tapajós could receive 42 schemes and reach a total capacity of more than 30,000 MW. The main schemes planned to start generating electricity by 2020 will be São Luiz do Tapajós (6,133 MW) and Jatobá (2,336 MW). One crucial difference when compared with other areas in the Amazon is that the Tapajós contains large plots of relatively undisturbed forests, several conservation reserves and indigenous reserves, all under threat by hydropower (Fig. 6.4). But several initiatives of the federal government demonstrate the perverse direction of its policies for the basin, as in the case of the law that reduced the territory of the Amazon National Park and the other 11 neighbouring parks (initially introduced as an emergency legislative act and then quickly transformed into ordinary legislation, the Law 12,678 of 2012) to facilitate the approval and licencing of the proposed hydropower projects. In 2014, a series of debates took place in the Tapajós to discuss the plans, but (as in the case of the Belo Monte dam) public consultation is highly controlled and does not leave much space for any critical assessment and consideration of alternatives. On the contrary, local communities and their allies have denounced the violation of the ILO Convention 169 that deals with the rights of indigenous and tribal peoples (Gazeta de Santarém 2013).

If in previous decades the Brazilian military deliberately tried to develop the region through large-scale, state-led economic production, the conversion of Brazil into one of the key emerging markets (epitomised as one of the BRICS countries) has fuelled a new 'rush for the north' involving both corporations and individuals. After

Fig. 6.4 Tapajós River at the extractive reserve Tapajós-Arapiuns

stabilising the economy according to neoliberal monetary policies in the 1990s, the federal government introduced the 'Forward Brazil' [*Avança Brasil*] development programme with several incentives to private entrepreneurs backed by the state. The more recent administrations, under the populist Presidents Lula and Dilma, replaced those strategies with similar initiatives specifically focused on the recovery of economic growth (under the suggestive title Growth Acceleration Programme—PAC). The programme included several infrastructure projects for the Amazon and the Tapajós, including paving the BR-163 highway (initiated in 2009 and is planned to be concluded in 2015, despite the fact that socio-environmental mitigation measures are lagging far behind engineering works). Lula also expanded existing conditional cash transfer schemes (conditional because they require beneficiaries to fulfil specified conditions in order to continue receiving grants), especially the 'Family Allowance' [*Bolsa Família*] that benefited millions of Brazilian families and helped his party (PT) to remain in power and win four general elections so far. In the communities of the lower Tapajós River, government alleviation schemes have increased the purchasing capacity of people who previously had no regular source of income.[3] On the other hand, it is possible to detect a clear criticism of the financial dependency and subtle discrimination promoted by the same programmes. Although the motto

[3]In 2011, another programme was introduced for those working in extractive activities in the Brazilian Amazon (called Green Stipend) and promises around US$ 150/month per family.

of Dilma Rousseff's administration is 'Rich Country is Country without Poverty' [*País Rico é País sem Pobreza*], action against poverty led by the federal government represents a series of short-lived mitigation schemes that fall short of questioning development trends and the legitimacy of huge social inequalities. As much as in the past, economic growth is taken for granted, reified as the inescapable imperative of a country that 'deserves' to be a new global power and where the Amazon cannot represent an 'obstacle' to national ambitions.

Minimum income schemes (i.e. means tested transfers) introduced by the federal government have not been immune to repeated criticism by other groups of the supposed indolence and inactivity of the local poor. Such (evidently unfair) blame of the poor hides the fact that such groups are those historically marginalised by the process of development and systematically exploited during colonisation, rubber production, agriculture expansion and urban expansion. In reality, the same social groups there were subjugated and benefited little from economic activities put forward for the Amazon by the main political and economic centres of Brazil are then exogenously labelled as poor in need of more of the same development, combined with end-of-the pipe poverty alleviation schemes, as in the case of the 'family allowance'. On the other hand, schemes like these may help to momentarily address poverty but in practice serve to legitimise and consolidate the overall pattern of aggressive and unequal development. In effect, state interventions to develop the Amazon have systematically undervalued the lives and the needs of traditional communities, peasants and indigenous groups, even when the alleged goals are to improve environmental conservation and legalise land ownership (Almeida 2011). The Sustainable Amazon Plan (PAS in Portuguese) launched by the Brazilian government in 2008 replicates old problems of centralisation, populism and market fundamentalism. In 2009, the national government introduced the programme Legal Land [*Terra Legal*] to regularise public areas illegally occupied in order to formally offset land grabbing and deforestation at the transition zone between the savanna [*cerrado*] and the forest. In practice, the programme served to undermine agrarian reforms elsewhere in Brazil and led to increasing land concentration and environmental degradation because it crated similar market-cantered mechanisms than those available for large-scale agribusiness (Oliveira 2013).

Overall, the Tapajós is now a decisive battlefield between, on the one side, the preservation of the Amazon as claimed by national and global NGOs and, on the other, powerful pressures for the expansion of large-scale hydropower generation, gold mining (especially gold panning) and agribusiness production (Osava 2013). In that challenging context, the governor of the state of Pará, Simão Jatene, publicly declared that the mistakes of past development seem to be once again replicated and, while the pressures constantly increase and the socio-spatial inequalities are magnified, there is no clear strategy for the Amazon and a growing gap between national and local authorities (interview at the TV Cultura's programme Roda Viva, 22 December 2014). In any case, while the state administration has recorded recent increases in life expectancy and GDP per capita in the Tapajós region, improvements in those aggregate indicators have been also associated with increases in the Gini index of social inequality from 0.488 to 0.499 between 2008 and 2011 (IDESP 2013).

More significantly, what has been happening in the Tapajós is an integral element of the advance of capitalism in the country, as a long, gradual reorganisation of the economy, society and the State, as described by the Brazilian sociologist Florestan Fernandes. The last author also argued that a 'bourgeois revolution' in the periphery of capitalism is 'essentially a political phenomenon' given the diversity of interests among capitalist groups (Fernandes 2005: 343). One of its concrete results is the widespread feeling of uneasiness about public policies and the devastating impact of private initiatives constantly encroaching upon new territories.

From the above, it is not difficult to see that in recent decades the Tapajós region was further incorporated into national development strategies as the source of exotic and valuable goods (including plant and mineral commodities), navigation and hydropower energy, as well as the deliberate deposit of poor people who could not find a better life in other parts of the country. The poor were both attracted to the area and then systematically contained through a combination of low-paid jobs, political repression, the use of forest resources for survivability and, increasingly, targeted poverty alleviation measures. Approaching it from a critical political ecology perspective, the region has been transformed, in highly contested ways, by both productive, profitable interventions and by the propagation of an economic model based on the twin processes of exploitation and inequality. In that sense, the insightful reflection of rural residents (characterised here as 'forest-dependent poor') helps to elucidate the obstacles to escape poverty and discuss the possibilities of change.

6.2 Poverty as a Mirror of Wider Socioecological Tensions

The trajectory of development in the Amazon, and in the Tapajós River Basin, in particular, raises several questions about the consequences of large-scale changes and the reinforcement of mechanisms of socio-natural exploitation that, in the end, produce new waves of poverty and marginalisation. Extensive areas in the Amazon have been converted into true landscapes of impoverishment where the prospects of a better life for the majority of the population are undermined by the very 'success' of economic growth (which is largely based on new extractivist activities controlled by powerful business groups). Poverty and affluence have evolved in interconnected ways through the transformations carried out in the Amazon in a manner that, as long denounced by Engels (1848), 'poverty and pauperism have been openly declared (…) to be necessary elements of the present industrial system and the national wealth'. Whereas the mainstream discourse considers wealth as the logical result of entrepreneurialism and efficient exploitation of territorial resources (especially mining, logging and farming), poverty effectively springs from corrupt relations of production and the systematic denial of justice (see George 1979). In official documents and mass media articles, the poor are typically located in a passive position, occupying spaces with clearly defined problems and in need of further integration with apparently thriving enterprises (e.g. The Economist 2013). Any interpretation of poverty which does not conform to the pre-established explanations and ongoing policies tends to be overlooked and subsequently removed from the public debate (Ioris 2012).

Poverty is a vast topic and certainly it cannot be exhausted here. Yet, based on qualitative research methods and engagement with locals (mainly in the RESEX area across the river from Santarém), two decisive lessons were learned and deserve special attention, respectively, about the disputed meaning of poverty in the Amazon and its persistence exactly due to the socioecological contradictions of development. Although only schematically considered in the present text, these two main findings help to inform a critical rethinking of poverty in the rich ecosystems of the Amazon. The *first lesson* starts with the recognition that, despite the fact that a multifaceted ontology of poverty has been well established in the critical academic literature, the intricate relationships between poverty making and the transformation of ecosystems under fast development pressures need yet to be better understood. It is even less common to come across analyses that interpret poverty metabolised through the appropriation and dealing with the forest in a way that incorporates socio-natural change into the resistance against social and political inequalities. The causes and consequences of poverty are rarely interrogated from the perspective of those living in close contact with forested ecosystems and witnessing their fast conversion into hotspots of economic development.

Interestingly, the very meaning of poverty for the locals differs from the official interpretation of national and international organisations. For instance, people in the RESEX demonstrate to be aware of major social differences but normally do not see themselves as poor, at least not according to the conventional account of poverty. They may not have had all the tools and goods of present-day society (as they can easily see on the television, which has a widespread audience), however their condition is one of basic needs satisfaction and, what is closely related, great collaboration across families and communities. As said in the interviews:

> We are poor because we are, but we have what we need. (senior resident, male, married)

> There are people with difficulties, but I don't think there is any poverty here. Everyone has a reasonable life. (young resident, male, single)

Poverty in the forest is therefore regarded as relative, lived and reacted upon through survival measures and the construction or retention of particular identities. It was also possible to uncover sophisticated and surprising elaborations on their condition and the opportunities provided by the interaction with the forest:

> Well poor, for me, poverty, for me, well there is poor and there is hard up. Today we are hard up, but we are not poor. A poor person has no home, no roof over his head, that's a poor person. I might not have anything today, but I will have tomorrow. Here at home it might have nothing in the morning, but by mid-day I'll have it. If I don't have what I need in the afternoon, I'll find a way get it by the evening. So we are not poor, the poor are those who don't have warm clothes, who don't have a source of income, who go around begging. That's what being poor is, we say, look at him poor thing, you don't say, look at that poor rich man. You feel sorry for him because he is poor. The word "poor" for me is very strong. (senior resident, male, single)

The poor are not simply people 'stuck' in the forest and needing help, but have dynamic lives, with a strong sense of place and able to construct vibrant socioecological relations (Campbell 2012). In many of our interviews, people stated that the forest offers opportunities and provide for, but that the help offered by the forest requires working, and working with, ecosystems (Figs. 6.5 and 6.6). Working the forest and with the forest are the main forms of getting by or escaping more acute conditions of deprivation. The material conditions of living are thus connected to specific forms of cultural knowledge, skills and social contacts valued by marginalised groups (Appadurai 2002; Robinson and Oppenheim 1998). Instead of the relatively easy discourse of environmentalists about protecting the forest, the environmental ethics of the poor is based on physical effort and appropriate knowledge:

> I used to work in the fields, then we came here and I worked on the fields here too, on the clearings. But then I decided to get into carpentry, I made a lot of canoes. (…) I think that [the forest] helps them escape from poverty. How? Working on the thing, you know, working, planting, reaping – then you have something. (retired, male)

This form of grassroots environmental ethics is put in practice through the constant and almost daily reworking of the forest, in a perennial practice that incorporates the condition of poverty into socio-natural relations. The forest ecosystems can provide some reassurance against the widespread sense of uneasiness about the rapid appropriation and conversion of the forest by large farmers, miners and engineering companies. That is articulated against the static and prejudiced conceptualisation of poverty by public agencies, which in many cases is instrumental in

Fig. 6.5 Honey production using bees native to the Amazon

Fig. 6.6 Women handicraft production making use of local plants

reducing the political role of the poor and their ability to modify their present situation. That is why alternative interpretations of the meaning and consequences of a poverty making geography, like those articulated by residents in the RESEX, offer a broader understanding of poverty as a complex set of practices and experiences which unfold through interrelationships between local cultures, opportunities and lifestyles (Alcock 2004; Danziger 2010; Room 1999).

However, even if the locals do not identify themselves with the ordinary characterisation of poverty, it does not mean that they don't perceive their inferior condition when compared with other rural and urban groups. But their protest is less about poverty in itself and more directly related to their role in the wider process of transformation and development (see also Barbier 2012). They react against the hegemonic tendency to convert highly valuable land resources to benefit mainly wealthy elites, thus exacerbating problems of inequality and discrimination while at the same time impeding economy-wide development:

> (…) we don't have everything we need to survive here. We have difficulties in reaching what we really want. But almost everyone here works and is able to support themselves with their work. (…) I think that for us who live a long way from the large capitals or towns, we depend on a lot of things that end up demotivating us to move forward because we have difficulties in these processes… (male, 35 years, married)

Local residents are not only fully aware of the value of their knowledge and the potential of their activities to overcome poverty but are also very clear on the many obstacles they inevitably face:

> Well, it didn't work because, like I said, the project comes, they work on it, but then we don't have the money to keep it going. Just so you have an idea, a bee-keeping project. We learned the technique, but when we had to start working we didn't even have the money to buy the boxes. Then there was the project for rustic furniture. Was it a good Project? Yes, it was. But afterwards, we tried to drag it on but you just can't do it like that. So that's what happens with the projects… (senior resident, male, single)

Most of the criticism is directed against the government and its close association, and convenient alliance, with the stronger players (timber, miners and agribusiness companies in particular):

> Because the government doesn't want to see where the poverty is. It only looks towards where there is more money. And we suffer the consequences here. (senior resident, male, married)

> For some of us the changes have brought good things, for others no. For example, for us who don't have the means to work with a lot of timber. A while ago there was a timber association here, there was plenty of wood around here, now there isn't. It can't be our main source of income because we don't have the machinery to cut much down. But for the big companies it brings a lot of money because they have the materials, the government helps them a lot and they have what they need. The weaker ones are the poor, the business man is not weak. (senior resident, male, married)

This leads us to the *second main lesson*, which is the genesis and prolongation of poverty located in the socioecological contradictions of development. The imbricate dialectics between wealth and poverty is mediated, and engraved on, mounting socioecological impacts resulting from the private appropriation of the commons. Instead of being simply anti-nature, the most harmful attribute of the process of development is its anticommons imperative, that is, the dominant model of development (based on the apparent and misleading 'abundance of nature') depends on the conversion of collectively owned and jointly managed tracts of the Amazon into the sphere of private property relations what is needed to allow the simultaneous exploitation of people and resources. Development is promoted as a remedy to poverty but in effect, it is simultaneously anti-poor and hostile to the commons. This disturbing convergence between anti-poverty and anti-poor tendencies is explained exactly because of a third, resulting pole of negativity, that is, the anticommons ontological basis of development. This happens in strike contrast to the largely communal world-view of the locals, for whom nature is the ultimate 'owner' of the land (as mentioned in an interview). Among the locals, the feeling that their hardship is connected with environmental degradation and labour exploitation due to the destructive intervention of new, powerful players is clear, for example, in the following extracts:

> We feel it for the forests. Not here, it doesn't happen much here, but where these big companies come into the forest we feel that, it may not be causing problems now, but we already feel that it could do later on. (37 years, male, married)

> In our region where it's just the little guy, the forest is not affected. But do you know who makes the situation worse? It's the big companies. They are the ones that are killing subsistence farming. They do everything in large quantities, they clear large areas and we are the ones that get into trouble. Here I can only clear four hectares, but the business man clears 200 [hectares]. Who degraded the forest more, him or me? And that's the thing. So I have just a little to eat, and he buys big car, he drives around and we are no foot, because I can't grow any more than this. I can't even buy a bull. There was this project from the government, they gave me a cart. It's still there today. And then one day one of them comes up to me and says, "So?", so I say to him, "It's over there. You give me a cart but you don't give me a bull to pull it!" Do you think I'm going to turn into a bull to pull the cart? That won't do. So that's it. (…) our agriculture doesn't affect the forest. (senior resident, male, married)

> (…) well, because the devastation of the forest…a lot of streams are drying up, because they destroy the forest. And the heat, why do we have that? It's very hot…. (retired, female)

Poverty is generated through the advance of mainstream development in the region and the reduction of socioecological complexities to the monodimensional sphere of market transactions. Once the commons were no longer protected, the stronger and more opportunistic competitors are able to exclude others from the access to now scarce, previously shared, resources. The implicit acceptance of those tensions makes sense in the long anticommons practice of development, as conceptualised in the post-World War II period, that typically associated it with the intensification of the production of goods and services according to the patterns of consumption (as well as waste) of Western societies. From being a remote land of exuberant biological formations that fascinated explorers for many centuries, Amazon was brought to the centre of national development policies that aimed to replicate similar technological, institutional and approaches. Contrasting with colonial and postcolonial times, when the riches were captured and removed, conventional development required the consolidation of private property inside the region and the reorganisation of social relations in function of the political power of private land, mines and industries (or similarly in function of state-owned property serving capitalist expansion, such as dams, roads and ports). The anticommons dynamics of Amazonian development is actually an expression of the wider phenomenon of alienation of humans from their product activity, social context and socio-natural condition. As argued by Marx (1988: 83), private property is the 'material, summary expression of alienated labour'; furthermore, private property embraces both the relation of the worker to work (and to the product of their labour) and the relation of the non-worker (the capitalist) to the worker and also the product of their labour.

The devastating impacts of the growth of private property institutions become evident in the growing erosion of subsistence, communal practices originally based on complex forest and ecosystem management. The risks associated with that were pointed out:

> Before, for example, fishing was easier. It was easy to fish around here, very easy. And today you don't see it because predatory fishing still goes on today. (…) Our water was very clean, crystal clear, today it's polluted. [This is also related to] the timber merchants that showed up (…) (retired, male)

Beyond the focus on efficiency by economists working on anticommons theory (e.g. Heller 1998), development as anticommons has codified inequality and served to disseminate poverty in the region. Socioecological transformations under the influence of development policies have unfolded through eco-class-race struggles that ultimately propagate and reinforce a poverty making geography (Isla 2009). Because of powerful anticommons pressures, the forest has been brutally altered (in symbolic and material terms) from its long-established, dynamic condition giving rise to both environmental disruption and social exploitation. Araghi (2009) observes that, according to Marx, it is the category of 'estranged labour'—i.e. the exploitation of workers for the accumulation of surplus value—that explains the structural dualism, and distancing, between society and nature.

It is evident that the main promoter and guarantor of anticommons trends has been the national state and its increasingly sophisticated combination of the development agenda and weak environmental policies contained by a vague idea of sustainability (Ioris 2014). The material consequence of strong anticommons ideology and associated practices is to leave the state in charge of the most decisive pressures on socioecological systems, at the same time that it has to negotiate the rate and the distribution of negative impacts. In other words, the hegemonic privatising force not only shapes development and produces poverty but also has consequences in terms of the distortions of environmental conservation policies. Supposedly innovative approaches to forest management adopted under the canon of environmental governance have likewise offered narrow, formalised solutions (e.g. carbon certification), which may be relevant to environmentalists and commercial land managers but are less relevant to the forest-dependent poor. Schemes that deal with environmental conservation tend to be top-down and highly prescriptive, rarely involving community participation at the local level and in meaningful ways (Ioris 2016). Such programmes are blighted by limited structure and coherence, and fraught with the lack of cross-institutional communication, while gaps in implementation and fragmented delivery aggravate deforestation and perpetuate poverty. The same governments that introduce environmental legislation and establish nature reserves constantly formulate economic incentives and construct roads and infrastructure that lead to further land concentration and aggravate conflicts. In practical terms, the anticommons commitments of the state, essential for the success and expansion of mainstream development, are inherently antagonistic to the conservation of forest ecosystems and the elimination of poverty-driven socioecological relations.

6.3 Resisting and Denouncing Development: Socio-natural Identity and Politics

The previous section discussed the context-specific meaning of poverty and the anticommons basis of development making use of the experiences and the reactions of communities living in the Tapajós basin. An examination of those two fundamental, and synergistically connected, processes help us to understand the politicised ecology of everyday life and the failure to fulfil demands that not easily fit in the mainstream agenda of economic growth and regional integration. According to this official agenda, development has to be promoted through intense resource exploitation and the 'productive' use of land, which have the perverse consequence of creating and maintaining multiple vulnerabilities. At the same time, efforts to alleviate poverty are, by and large, hampered by an overly simplistic representation of economic development and of the multiple scales across which drivers of poverty and environmental degradation operate. While some government interventions have brought positive but localised results, as a whole, investments and assistance programmes have failed to produce the desired recognition of wider social rights and the correspondent valorisation of the socio-natural features of the Amazon. The poor are stereotyped and assumed to be culturally backward and incapable of escaping poverty on their own.

Beyond any mainstream consideration of the causes and remedies for poverty, the deprived condition of rural communities is really the outcome of a powerful hegemony that has been applied simultaneously to both of them and the forest. This prevailing direction of economic development does not diminish the value people attach to community life and their crucial connections with the land. On the contrary, the impasse of development and poverty making will only be overcome when substantive solutions that can be found through contextual, place-based approaches to resources and socio-natural relations. Reworking the forest and with the forest—for example, producing subsistence food, artisanal artefacts and other objects that can be sold in local or national markets—constitutes an element of the concrete alternative to the perverse anticommons trends of mainstream development. The results of our research directly indicate that communities in the Amazon cleverly associate, in a highly politicised way, the value of the forest with the value of their own labour, in a way that all spheres of value (intrinsic, use and exchange value) are inextricably linked (see also Kovel 2014, for the importance of a critical epistemology of values).

As a concluding point, it is relevant to observe that most scholars have examined the poor according to their own intellectual biases and academic commitments, allowing the poor only an instrumental role as holders of limited political agency. It was denounced by Rancière (2004: 81) that even radical authors when dealing with class consciousness, reduce such groups simply to their revolutionary duty, what in practice 'is nothing else than the negation of the worker'. The philosopher addresses the apparent dilemma of the needed transformation of the world being pursued exactly by a small group of people who prefigure, fight for and, ultimately, deserve a better life. But Rancière only treats this impasse as a purely politico-economic problem, while in reality the main challenge is profoundly socio-natural and politico-ecological. The main argument here is that the agency of the poor (or the proletariat) is not based on their destitution or class identity, but derives primarily from their socioecological identity and vital interconnections with the rest of nature. Equally, it will never be possible to overcome poverty without confronting the hegemonic forces that persistently undervalue the socio-natural whole and accumulates a capital from the deliberate fragmentation and exploration of socio-nature.

Acknowledgements This chapter is based on the article: Antonio A. R. Ioris (2015). The Production of Poverty and the Poverty of Production in the Amazon: Reflections from Those at the Sharp End of Development, Capitalism Nature Socialism, 26:4, 176–192, https://doi.org/10.1080/10455752.2015.1058835. Reprinted with the kind permission of Taylor and Francis, and The Center for Political Ecology.

References

Alcock P (2004) The influence of Dynamic perspectives on poverty analysis and anti-poverty policy in the UK. J Soc Policy 33(3):395–416

Almeida AWB (2011) Traditionally occupied lands in Brazil (trans: Mathews M). UFAM, Manaus

Appadurai A (2002) Deep democracy: urban governmentality and the horizon of politics. Public Cult 14(3):21–27

Araghi F (2009) Accumulation by displacement: global enclosures, food crisis, and the ecological contradictions of capitalism. Review 32(1):113–46

Barbier EB (2012) Corruption, poverty and tropical land use. J Sustain For 31(4–5):319–339

Bunker SG (1985) Underdeveloping the Amazon: extraction, unequal exchange, and the failure of the modern state. Chicago University Press, Chicago and London

Campbell JM (2012) Brazil's deferred highway: mobility, development, and anticipating the state in Amazonia. Boletín de Antropología 27(44):102–126

Castree N (2002) False antitheses? Marxism, nature and actor-networks. Antipode 34(1):111–146

Castro E (2007) Políticas de Ordenamento Territorial, Desmatamento e Dinâmicas de Fronteira. Novos Cadernos NAEA 10(2):105–126

Castro E (ed) (2008) Sociedade, Território e Conflitos: BR-163 em Questão. Belém, NAEA

Costa SMG (2012) Grãos na Floresta: Estratégia Expansionista do Agronegócio na Amazônia. PhD thesis, NAEA/UFPA, Belém

Danziger SK (2010) The decline of cash welfare and implications for social policy and poverty. Ann Rev Sociol 36:523–545

Dean W (1987) Brazil and the struggle for rubber: a study in environmental history. Cambridge University Press, Cambridge

Engels F (1848) The Kölnische-Zeitung on the state of affairs in England. Neue Rheinische Zeitung 62(1). https://www.marxists.org/archive/marx/works/1848/08/01.htm

Fernandes F (2005) A Revolução Burguesa no Brasil: Ensaio de Interpretação Sociológica, 5th edn. São Paulo, Globo

Freitas M, Freitas MCS, Ioris AAR, Castro WE Jr (2017) Amazônia. Chiado Editora, Lisbon

Gazeta de Santarém (2013) Comunidades Querem ser Ouvidas sobre Usinas Hidrelétricas no rio Tapajós. http://www.gazetadesantarem.com.br/regional/comunidades-querem-ser-ouvidas-sobre-usinas-hidreletricas-no-rio-tapajos

George H (1979[1879]) Progress and poverty. Robert Schalkenbach Foundation, New York

Gondin N (2007) A Invenção da Amazônia, 2nd edn. Manaus, Editora Valer

Heller MA (1998) The tragedy of the anticommons: property in the transition from Marx to markets. Harv Law Rev 111:621–688

Hecht SB (2013) The scramble for the Amazon and the lost paradise of Euclides da Cunha. University of Chicago Press, Chicago and London

Holanda SB (2000) Visão do Paraíso. São Paulo, Brasiliense

IDESP (Pará's Institute of Economic, Social and Environmental Development) (2013) Mapa de Exclusão Social do Estado do Pará. Belém, Pará State Government

Ioris AAR (2007) The troubled waters of Brazil: nature commodification and social exclusion. Capital Nat Soc 18(1):28–50

Ioris AAR (2012) The persistent water problems of Lima, Peru: neoliberalism, institutional failures and social inequalities. Singap J Trop Geogr 33(3):335–350
Ioris AAR (2014) Environmental governance at the core of statecraft: unresolved questions and inbuilt tensions. Geogr Comp 8(9):641–652
Ioris AAR (2016) The paradox of poverty in rich ecosystems: impoverishment and development in the Amazon of Brazil and bolivia. Geogr J 182(2):178–189
Ioris AAR (2018) Centralidade da Fronteira: Ensaio sobre a Origem e Evolução de Fronteiras Sócio-Espaciais. Territórios e Fronteiras 11(2):23–41
Isla A (2009) The eco-class-race struggles in the Peruvian Amazon Basin: an ecofeminist perspective. Capital Nat Soc 20(3):21–48
Kindon S, Pain R, Kesby M (eds) (2010) Participatory action research: approaches and methods. Routledge, London and New York
Kovel J (2014) Ecosocialism as a human phenomenon. Capital Nat Soc 25(1):10–23
London M, Kelly B (2007) The last forest: the Amazon in the age of globalisation. Random House, New York
Marx K (1956[1847]) The poverty of philosophy. Lawrence and Wishart (for the Foreign Languages Publishing House), Moscow
Marx K (1976[1867]) Capital: a critique of political economy, vol 1 (trans: Fowkes B). Penguin, London
Marx K (1988[1844]) Economic and philosophic manuscripts and the communist manifesto (trans: Milligan M). Prometheus, Amherst, NY
Oliveira GLT (2013) Land regularization in Brazil and the global land grab. Dev Chang 44(2):261–283
Ollman B (1976) Alienation: Marx's conception of man in capitalist society, 2nd edn. Cambridge University Press, Cambridge
Osava M (2013) Tapajós, a Batalha Decisiva pela Amazônia Pode-se Negociar. Envolverde, 26 Dec 2018. http://envolverde.cartacapital.com.br/tapajos-a-batalha-decisiva-pela-amazonia-pode-se-negociar
Pereira JCM, Leite MSP (2011) A "Fala do Desenvolvimento" em Belterra e a Transformação do Lugar em Dois Contextos de Modernização. Novos Cadernos NAEA 14(2):197–217
Pinto LF (1980) Amazônia: No Rastro do Saque. Hucitec, São Paulo
Pogge T (2008) World poverty and human rights, 2nd edn. Polity Press, Cambridge
Rancière J (2004) The philosopher and his poor. Duke University Press, Durham and London
Rivero S, Jayme FG Jr (eds) (2008) As Amazônias do Século XXI. Belém, UFPA
Robinson P, Oppenheim C (1998) Social exclusion indicators: a submission to the social exclusion unit. Institute for Public Policy Research, London
Room GJ (1999) Social exclusion, solidarity and the challenge of globalization. Int J Soc Welf 8(3):166–174
Santos BS (2014) Epistemologies of the South: justice against epistemicide. Paradigm Publishers, Boulder and London
Silva MC (2012) O Paiz do Amazonas, 3rd edn. Editora Valer, Manaus
The Economist (2013) Poverty, geography and the double dilemma, 1 June 2013. http://www.economist.com/blogs/feastandfamine/2013/06/aid-agencies-future
Zaitch D, van Solinge TB, Müller G (2014) A green criminological and human rights perspective on land-use change. In: Bavinck M, Pellegrini L, Mostert E (eds) Conflicts over natural resources in the global south conceptual approaches. CRC Press, Boca Raton, FL, pp 91–108

Chapter 7
Disrupting Frontier Development from Within: The Latent Geographical Agency of Indigenous Peoples

...the Indian problem is us,.
who invaded their lands and destroyed their lives.
Darcy Ribeiro (2015)

Indian is not the problem, Indian is the solution.
Anastásio Peralta (Kaiowa), Panambizinho Reserve, in an interview

Abstract The chapter examines the survival strategies and political reactions of indigenous groups in areas of agricultural frontiers that are informed by cultural symbols, family bonds and land-based responses. It discusses the unique socio-spatial trajectory of indigenous peoples and, in addition, proposes a typology of indigenous spaces. The analysis is focused on the emblematic example of how frontier making was experienced by the Kaiowa-Guarani of South America. The wisdom and resistance of Kaiowa-Guarani groups derive from the simultaneous ethnicisation of space and spatialisation of culture. They have shown latent geographical agency shaped by religious practices, strong family ties and the ability to internally negotiate the return to their original areas. There are many lessons to be learned, in particular, the talent to absorb the increasing and dissimulated brutality of frontier making and, at the same time, voice their political demands, form solid strategic alliances and coordinate land recovery initiatives.

Keywords Guarani · Kaiowa · Land-grabbing · Retomada · Culture · Religion

7.1 Spiralling Paradoxes: The Indigeneity of Frontier Making

As we have been arguing throughout this book, socio-spatial frontiers are not just remote areas where new socio-economic relations are imposed on top of previous conditions. The space of the frontier has a complex, multifaceted configuration in which both the hyper-new and what is considered obsolete coalesce. Following development and modernisation trends, frontier making brings the potentiality and, more

A. A. R. Ioris, *Frontier Making in the Amazon*, Key Challenges in Geography,
https://doi.org/10.1007/978-3-030-38524-8_7

importantly, the long-lasting problems of core spatial areas to the emerging frontier spaces (Ioris 2018a). This involves the dislocation, incorporation and exploitation of an existing state of affairs translated into novel opportunities to succeed, earn and accumulate. Among the many peoples affected, indigenous groups—also described as natives, ancestral, first nations, Fourth World or aborigines—have vividly experienced the complex interplay of clashes, violence and resistance due to their brutal insertion in the process of frontier making. Discourses of sovereignty and legitimacy certainly underwrote 'settler colonialism's desire to uproot and destroy the place-based autonomies of Indigenous peoples in the relentless acquisition of ever more land and resources (Larsen and Johnson 2017: 4). The colonial project was a coordinated attempt to affirm the power of the European nation states and create sources of wealth and economic rents out of the subjugation or displacement of indigenous peoples, with survivors normally forced to seek refuge in remote areas, already occupied by other groups, and which would later be incorporated again as new frontiers. Particularly in Latin America, the tragic history of indigenous groups is well known, written under the auspices of colonisation and following the pattern of a racialised power exerted both by the European powers and by governments in Latin America after independence (Quijano 2008).

Mariátegui (2007), almost a century ago, rightly denounced the idea that the 'Indian question' in the region was primarily a problem of the land ownership regime. However, the process of frontier making was a devastating blow that went far beyond only land or resource grabbing: it has been a true phenomenon of world-grabbing in the sense that it has been an attempt to reduce lives and landscapes to the language of money and profit. Unlike migrant peasants, miners, construction workers or farmers, indigenous peoples had already been living in the areas converted into frontiers, which raises an important question regarding how they differ from other subaltern groups involved in frontier making activity. Martins (1995) argues, based on studies in the Amazon, that the condition of squatters at the frontier results directly from the contradictions of capital, given that squatters must be removed from disputed areas in order for private property and wage labour to flourish. For that reason, the 'squatter is product of the very logic of the expansion of capital, whilst the Indian is not [such a product]. The squatter can be reimbursed for his work, as a means to remove him from the land pursued by the farm or the company… The Indian cannot, in principle, be reimbursed to leave his land: this land is not just a thing, an object… it doesn't have the attribute of thing and commodity. The land is sacred, it is the base of the tribal organisation' (p. 117). On the other hand, the political condition of indigenous peoples is not entirely different and, just like peasants, squatters and similarly marginalised people, their existence is intensely transformed by their immersion in frontier making. The dialectical result is that indigenous groups retain their cultural heritage and strict relations with the land, but at the same time they share important patterns of mistreatment and marginalisation with other groups. As pointed out by Anthias (2017), indigenous territories are neither separate from, nor entirely subsumed by, capitalist development processes, but rather subject to multiple land values, ontologies and business investments.

Therefore, the participation of indigenous peoples in frontier making, although widely studied throughout the world, still demands further conceptual and empirical scrutiny. The typical academic narrative, influenced by a colonialist mindset, describes the waves of domestic or international migration promoted by national or international settler states to foster economic activity, at the expense of the land and resources of those who were already living in the area. These accounts normally deal with socio-spatial control, and in some cases genocide, historically exerted to crush political opposition. The focus in such texts is on the systematic imposition of a new socio-spatial order by the settler states on indigenous peoples, at the expense of pre-existing institutions such as common ownership of land, a self-sufficient economy, nomadic stateless life and spiritual bonds to the land (Yiftachel and Fenster 1997). They discuss how the original inhabitants of the territory were obliged to retreat in the face of the advance of the new frontier, forcing them into situations of acute exploitation, deprivation and overcrowding. However, this literature has frequently overlooked the perceptions, the inventiveness, the active reactions and the complex ontology of indigenous peoples. It is really only since the 1970s that indigenous voices and political subjectivity have started to attract better recognition. Indigenous groups are the quintessential targets of resource colonialism and the theft of land and resources (Parson and Ray 2018), but their history and agency did not end with the loss of their land and decimation of members of their society. On the contrary, they continue to claim an indigenous identity in daily life activities and maintain attachments to places under difficult circumstances. Consequently, the indigeneity of frontier making requires postcolonial sensibilities and recognition of the political significance of culture in space. Beyond essentialist and reductionist positions, there should be a concern for the politics of race, as well as questions of representation and the ideological construction of various racialised 'others', in favour of conceptualisations that are time and place specific (Jackson and Penrose 1993).

Our goal is not to review in any great detail the geography, culture and history of indigenous groups, but to examine their survival strategies and insurgent geographies based on cultural symbols, family bonds and land-based responses to the excesses of frontier making. This is directly inspired by the 'cultural materialism' of Raymond Williams, which entails the mobilisation of culture against modern society with a creative reconciliation between multiple sensitivities and an anti-capitalist critique (Löwy and Sayre 2018). The main intention is to make use of the conceptualisation of frontier making to understand the socio-spatial trajectory and, in addition, propose a typology of indigenous spaces. This chapter will concentrate on the emblematic example of how frontier making was experienced by the Kaiowa, one of the Guarani peoples of South America (for background information, refer among others to Brand 1997; Chamorro 1998; Clastres 1974; Melià 2004; Pereira 2016). It is estimated that the Guarani population comprises around 280,000 individuals, of whom 85,000 live in Brazil, 80,000 in Bolivia, 61,000 in Paraguay and 54,000 in Argentina (EMGC 2016). They are normally described as belonging to four clustering groups: Kaiowa (the denomination used in Brazil, also known as Paĩ-Tavyterã in Paraguay), Nhandeva (also described as Ava-Guarani), Mbya (in the south of Brazil, Paraguay and Argentina) and Chiriguanos (at the foothills of the Andes in Bolivia). The majority of the Kaiowa-Guarani (around 45,000 people) live in the Brazilian state of Southern Mato Grosso [*Mato Grosso do Sul*] around and to the south of the city of Dourados, often sharing reserves and lands with other ethnic groups (such

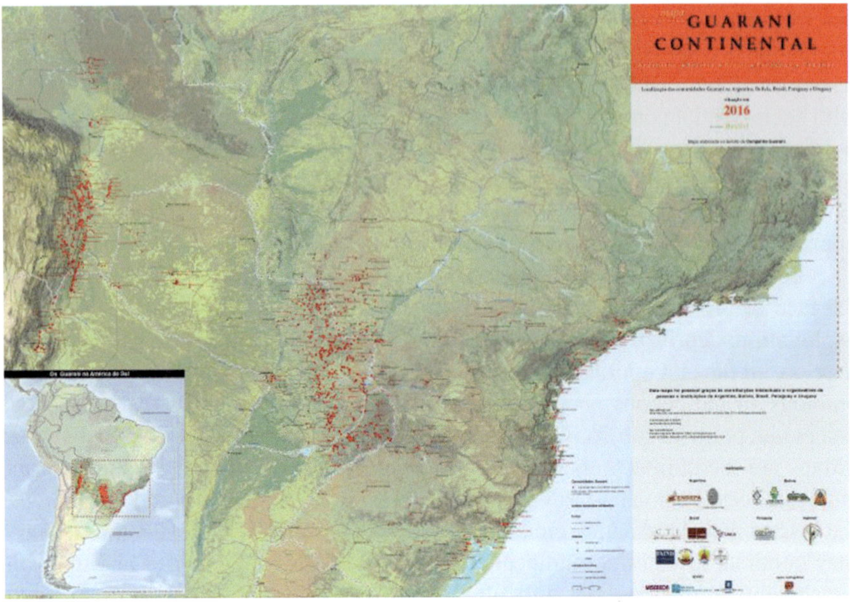

Fig. 7.1 Present-day distribution of the Guarani population in South America (*Source* EMGC 2016)

as Nhandeva-Guarani and Terena).[1] Figure 7.1 shows the population distribution of the Guarani peoples of South America.Disruption and aggression towards Kaiowa communities is a long-established phenomenon, originally related to the process of colonisation and aggravated during moments of territorial dispute, such as during the Paraguay War (1864–1870), the bloodiest conflict in the history of the continent, which took place right in the middle of Guarani territories. From the end of the nineteenth century, after the war, explorers and companies were attracted to the region due to its abundant natural resources, good agricultural land and strategic location in terms of river navigation, relative proximity to the main economic centres and the geopolitical need to occupy international border areas (Ioris 2012). Economic activity has intensified in recent decades due to the expansion of export-oriented agribusiness and increasing land speculation. One crucial driving force behind the various periods of colonisation and economic production has been the distressing levels of violence against individuals and families, particularly in areas of acute agrarian conflict between farmers and indigenous collectives. Space-based violence is a major, ongoing reality, not a singular event that happened in the past, that is, the oppression of indigeneity continues to pervade the geographical present (Radcliffe 2017). Indigenous groups are systematically harassed by the police and by private

[1]Although not formally in the Amazon Region, Southern Mato Grosso has been affected by the same national policies and politico-economic processes of frontier making, hence its inclusion as a relevant case study.

militias hired by landowners (known as *pistoleiros* i.e. 'hired guns'). Professional militias are typically registered as private security firms, managed by former military police officers and tacitly accepted by a pliant judiciary. Even in the regularised indigenous reserves, the population feels discriminated against and marginalised by the rest of society, leading to growing frustration and social unrest. Kaiowa-Guarani communities have presented alarming rates of suicide (555 between 2001 and 2011, according to CIMI)[2] and premature deaths (including children hit by cars) are much higher among indigenous families than in the rest of society.

Out of the total number (1,119) of murders of indigenous individuals in Brazil between 2003 and 2017, 461 cases (41.19%) occurred in Southern Mato Grosso (CIMI 2018a), particularly affecting the Kaiowa and involving atrocities that amount to genocide (Zelic 2018).[3] Such macabre statistics are related to overcrowded reserves, chronic malnutrition, usurpation of ancestral lands and disregard or attacks from agribusiness farmers, mainstream politicians, authorities and the business community. During our research in Southern Mato Grosso, discrimination against Kaiowa children and adults was disturbingly common in shops and streets, in addition to mistreatment by civil servants. Court cases often take decades to be decided, even when the rights of the indigenous groups are clearly defined and easy to ascertain. More than economic, political and environmental refugees, the Kaiowa are expatriated in their own territory, as they have been systematically marginalised from the benefits of development and forced to live in newly created, and constantly reinforced socio-spatial edges. Despite using social media to promote their cause, the Kaiowa are much less visible, for example, than national and international NGOs because of internet algorithmic filtering, which deliberately excludes some types of political discourse, segregates social groups and suppresses divergent perspectives (Ochigame and Holston 2016). Racism is concrete, structural and widespread (despite the fact that it is officially a serious, crime according to national legislation) and, in that regard, the Kaiowa-Guarani share many of the troubles affecting other indigenous populations, peasants and the urban poor throughout the country.

The shocking poverty and terrible suffering of the Kaiowa-Guarani represent the unpleasant side of the agribusiness-based regional economy and corrodes the supposed legitimacy of development and the existence of the democratic rule of law in the region. Agribusiness causes a range of problems for the world at large, but at the frontier it is even more belligerent and irrational, compensating and filling state and civil society gaps with the rhetoric of 'development' when in reality the region has been launched into a state of low-intensity agrarian warfare between farmers and indigenous groups. Nonetheless, the spatial condition of the Kaiowa is

---

[2]CIMI is the Indigenist Missionary Council founded in 1972 by the Catholic bishops of Brazil to defend the rights of Brazilian Amerindians.

[3]Today Brazil has 305 indigenous peoples who speak 274 different languages (there were 5,000 groups in 1500, the year of the Portuguese conquest) and occupy about 13% of the national territory (although 97% of this area is in the Amazon). Southern Mato Grosso has the second largest indigenous population, but the worst allocation of land in the country; the population density is 10.18 inhabitants/km^2, which is 40% more dense than among the non-indigenous population, and in the case of the Kaiowa it reaches 34 inhabitants/km^2 according to data presented by Zelic (2018).

a true riddle for regional economic planning: on the one hand, their land, resources and labour force are central to the very success of the frontier; on the other, the existence of indigenous groups means that the worst side of frontier making is still active and constantly reinforced by the unresolved incongruity between cultures and cosmologies. The result is that indigenous groups such as the Kaiowa are the embodiment of the spatial frontier and also its most explicit subversion. The Kaiowa-Guarani articulate a resistance against political hegemony, an ability to operate in the interstices of the state apparatus and a capacity to attract support from a heterogeneous array of organisations; this can be described as 'latent geographical agency'. Before examining the specific situation of the frontiers lived and transgressed by the Kaiowa, it is important to reflect upon the multifaceted condition of indigenous peoples and the related challenges involved in conducting research from a critical and ethical perspective.

7.2 Making Sense of the Indigeneity of Frontier Making

In this section, we ponder how to properly consider the deeply politicised condition of indigenous groups at the frontiers of economic development. Our reflection is closely associated with the emerging 'critical geographies of indigeneity' (Radcliffe 2017), meaning, in particular, an appreciation of landscape as a framework for addressing basic human rights, that is, the role of a 'right to the landscape' in the movement towards justice, dignity and well-being, integrating the spiritual and cultural values of land and local communities (Egoz et al. 2011). The landscape becomes part of what people are through everyday life experiences, and indigenous people create meaningful relationships and connections with the land as their main source of survival (Ingold 1993). To fully grasp that, it is necessary to acknowledge the long positivistic tradition and colonial foundations of Western knowledge, going back to the nineteenth century and even before, typically focused on areas of high biodiversity and abundant natural resources (MacDonald 2017). As pointed out by Coombes et al. (2014: 845), working with indigenous peoples has stretched 'presumptions about appropriate modes of engagement and representation… that challenge reaches the heart of the enterprise to question the very purpose of research.' Among other methodological and epistemological consequences, this means conducting research with and for the benefit of indigenous communities, allowing the expression of their own voices and their involvement in the interpretation of findings, rather than the conventional research on those communities for the benefit of non-indigenous scholars and government agencies. In addition, and similar to feminist, anti-racist and queer movements, indigenous geography opens the prospect of decolonising and reimagining wider horizons and functions for/of geography (Panelli 2008). As MacDonald (2017) splendidly observes, the research process is even more important than the immediate outcomes.

Many social scientists have demonstrated a growing commitment to embrace reflexive methods and deal with the politics of representation and the relational basis of critical research (Smithers Graeme and Mandawe 2017). This represents a challenging, but rewarding attempt to remove prejudices and learn together through, for instance, participatory research within fluid situations fraught with injustices and ongoing territorial disputes. Several relevant aspects deserve particular attention, starting with ethical questions. Despite the debate and the concerns about research ethics, there still remain important barriers that continue to limit the understanding of indigenous landscapes affected by the pressures of economic development. Among the thousands of Kaiowa in Southern Mato Grosso, the majority live in the congested reserves and around a quarter live in urban areas or in encampments along the roads. This is a very unsettled and painful situation that evolves through survival strategies amid threats of assimilation, annexation and extinction. Sociocultural information about the Kaiowa is, therefore, deeply contingent upon specific space making circumstances, exacerbated by recurrent conflicts, the need to maintain family networks and external alliances and the challenge to make sense of, and survive, in-between world situations. The Kaiowa-Guarani have skilfully mobilised their memories, imaginaries and cultural expressions towards acts of resistance and spatial practice, and their involvement in any research project is seen as part of the struggle and a search for allies. As a result, researchers find themselves 'in-between worlds' and are irrevocably transformed by the experience (Larsen and Johnson 2012). Researchers deal with accumulated tensions and passionate discourses, which often bring them to centre of the struggle. It is a movement away from representing the Other and towards collective problem-solving, activism and advocacy (Coombes et al. 2014). The researcher occupies a specific location within the broad sociopolitical context of their research and must responsibly handle how this positionality and associated privileges shape knowledge construction (Smithers Graeme and Mandawe 2017).

Ethical concerns must be sensitive and creatively engage with the past trajectory, the current political situation and the aspirations of societies in the midst of land-based disputes with uncertain prospects. This is clearly important in terms of the confidentiality of social and personal information, and the constant risk of revealing details about leaders and their strategies, which could undermine action or influence public perception. It could likewise worsen internal divisions among indigenous families or groups, as well as between them and their traditional allies (churches, unions, NGOs, etc.). Riddell et al. (2017) list a number of crucial requirements for conducting ethical research, such as the informed and autonomous engagement of indigenous participants, recognising their ownership, control, assessment and possession of knowledge, respecting their intangible cultural property (i.e. language and traditions), reciprocity and inter-relational accountability. It means that the study of indigenous life cannot be contained within the narrow boundaries of non-indigenous science and reasoning, as much as the basic terminology (i.e. indigeneity, aboriginality, colonial dominance, etc.) is highly contested. In the case of the Kaiowa, it is also necessary to consider their acute political situation and how they mobilise culture and religion to recover lost landscapes. What is needed is more than the usual participatory action research, but rather an engaged research approach that recognises local systems of knowledge and practice as fully authoritative and the actors involved as sufficiently competent to design, conduct and evaluate the research they are involved in. This is to avoid what is happening with supposedly inclusive techniques, such as participatory mapping, that are being incorporated into policymaking by government agencies and multilateral organisations. The indigenous participants should be more than informants or collaborators but should be treated as co-researchers, co-ethnologists, co-creators.

The researcher must establish trusted connections with real people, without romanticising events or political leaders, that is, avoid dealing with the 'hyperreal Indian' of many NGOs, a fantasy that reinforces the simulacrum image of indigenous people, supposedly pure, ecological, stoic, unadulterated (Ramos 1994). Anthropological research has specifically demonstrated that the notion of the Kaiowa person emerges from relations across wider categories of their society, it is an intense dialectic between the self and their collective condition (hence their discrimination against single adult males). Among other issues, it is problematic to think of clearly distinct epistemological differences between approaches loosely identified as 'Western' and 'indigenous', because in effect these are both polyvalent and polyvocal realms of discourse, just as indigenous people live within the framework of Western cultures (Shaw et al. 2006). Indigenous cultures and identities are fluid, their narratives and engagement with place and space are mutable, not linear; all this invites and prompts experimentation, innovation, affection and partnerships (Coombes et al. 2011; Murton 2012; Ramos 2018). Related to that, it is unacceptable for the Kaiowa that researchers would not give clear feedback and yield some concrete benefit back to them in relation to their struggle for recognition, rights and land. As pointed out by an anthropologist in an interview, the research effort needs to recognise that the political demands of indigenous peoples are complex, multiple and constantly changing because of emerging problems. The responsibility of the researcher is to conduct objective research, to the best of their capacity, while at the same time responding to the expectation of a supportive and worthwhile interaction. The geographical trajectory of the Kaiowa encapsulates disputes and reactions that are common to other indigenous groups and even peasant communities throughout South America, but they have also experienced unique challenges related to their location, specific geographical settings and particular involvement in the wider process of modernisation through regional development.

While they expect support from researchers, the Kaiowa are not desperately begging for help from non-indigenous actors. On the contrary, and in common with other contexts where extreme oppression has led to mobilisation, they have a lot to contribute in terms of resistance and ability to react. Despite sustained violence over almost two centuries, with different stages of land-grabbing and cultural obliteration, the Kaiowa have resisted and their population is on the rise. They represent an uncomfortable presence in the process of regional development based on agribusiness intensification and dominated by large private properties. The fact that indigenous groups even exist in the region, against all the odds, suggests that the sociocultural landscape of Southern Mato Grosso is much more complex than normally considered in reductionist analyses that reduce interactions to forces in favour of or against development. In fact, the survival of the Kaiowa (and the other first nation groups in South America) reveals their remarkable strength and the astute strategies they have adopted (obviously, in no way detracting from their pain or minimising significance of the long list of assassinations and attacks against them). There are elements in their culture and value system that are not easy to capture and bring to the realm of academic assessment, but it is not too difficult to understand that their cosmovision, their religious beliefs and practices and their linguistic and artistic skills come together to form a coherent wisdom. As argued by the late Kaiowa leader Ambrósio Vilharba, retaking land is not the end in itself, but it 'will reinforce our way of being Kaiowa' (in Markus 2013: 9). This sensitivity results from the accumulation of long-term collective engagement with the landscape, reaction or interaction with non-indigenous groups and the state, and the ability to incorporate and metabolise elements of the new world order. The tortuous, fragmented process of resistance and reaction is, essentially, a perennial attempt to project indigeneity into a space that is itself highly dynamic. The Kaiowa have had to struggle and wait, but—notwithstanding all the troubles and recurrent pain—have kept alive the possibility (often translated into reality) of restoring spatial indigenisation. The irony is that this is ultimately an effort to re-indigenise a territory that used to be theirs.

A transformative indigenous research effort needs, therefore, to recognise the limitations of academic work (particularly in the present neoliberal context where universities are increasingly managed as businesses and academics are seen as money-making labourers) and dislocate the centre of knowledge production: knowledge is creatively produced on the ground, through an open, horizontal engagement between different voices in which the academic researcher is basically a facilitator, a translator of what has been jointly learned. Indigenous co-researchers should be able to have control over collecting information about themselves, to access and analyse information according to their own needs and goals, determining what should be communicated and how. At the same time, research practice must be planned to adequately comprehend, respond and engage with social groups facing constant threats. Considering the context of violence and the willingness of the Kaiowa to pay a high human price in order to get their lands back, the research effort should be extremely careful to neither undermine ongoing mobilisations nor put people and leaders at risk (our visits to road encampments and *retomada* areas were certainly moments of high tension and apprehension for both the indigenous families and other research colleagues). This is not a trivial logistical question, given the recent history of the Kaiowa, marked by record numbers of murders and suicides. Likewise, there is a clear need to theorise the world from the perspective of the indigenous groups, rethinking universal concepts and searching for alternative socio-economic and political paths. Indigenous settings and cultures continue to require appropriate conceptual, methodological and interpretative approaches. Their contemporary condition is not only shaped by constant attacks and the pain of losing their land, but also by the resolve to resist and react, which will be analysed below.

7.3 A Frontier Making Avalanche …

Many readers are likely to have heard about the Guarani of South America in contexts including, for example, narratives about their imposing architecture and a complex society managed by Jesuit priests in the centre of the continent during the colonial period. Some may also be informed about the ongoing violence against the Kaiowa of the state of Southern Mato Grosso (considered the Gaza Strip of Brazil). But few will be fully aware of the scale of land rights violations, the systematic killing of adults and children, ferocious discrimination against people living in precarious settlements along the roads or in the periphery of the cities, widespread suicides of young people and teenagers, and disturbing levels of insalubrity and food insecurity. The Kaiowa have been exposed to the worst side of frontier making, affected by land-grabbing, displaced from their ancestral lands, confined to minuscule reserves and have experienced widespread socioecological exploitation (ranging from underpaid work to deforestation of their reserves for timber extraction). Kaiowa-Guarani life based on common land managed by large families and regular movement around the region has been (partly) substituted by the commodification of labour, land and nature, and, increasingly, urbanisation, drug trafficking and cross-border contraband. There is actually no other way to describe the situation of the Kaiowa under the advance of soybean and sugarcane production (among other crops) than genocide, following Short's (2016) definition which includes cultural destruction, social death and ecological devastation as equally important genocidal practices.[4]

It is crucial to observe that the advance of the economic frontier over indigenous land in Brazil was not only violent and disruptive but primarily illegal. First and foremost, the Kaiowa have been the lawful owners of their land since colonial times and according to the rules imposed on them. In 1680, a Royal Decree by the King of Portugal instituted the *indigenato*, which guaranteed that the concession of land to private individuals should not interfere with the 'original rights' of the indigenous groups to their land, as well as exempting them from any duty or tribute. This old legal principle that emanated from Portugal was gradually incorporated into Brazilian legislation, although in practice it was very easy to bypass the law and obtain land from the state or by employing widespread fraud (Vietta 2013). The legal system clearly guaranteed the rights of indigenous peoples and the protection of their cultural and material needs, but that did not prevent the concession, after 1880, of millions of hectares of Kaiowa-Guarani land to the Matte Larangeira company, a corporation that profited from the native mate herb (*erva-mate*) through the exploitation and

[4]Much of the debate on the impacts and legacies of frontiers revolved around whether or not many indigenous groups were subjected to genocide and genocidal practices (Rogers and Bain 2016). However, van Krieken (2004) insists that the source of destruction may lie less in an 'unambiguous "intent to destroy" a human group, than in the presumption that there was not much *to* destroy', which means that the colonial settler ethos, although it may not spell out a clear plan for destruction, certainly provides ideational sponsorship for genocidal actions (quoted in Woolford and Thomas 2017: 71). Rather than focusing on legalistic notions of 'intent' by settlers and governments, it is decisive to acknowledge the destructive potential of the systematic practices of cultural devaluation and social marginalisation..

even the scandalous, but tolerated, enslavement of indigenous populations. A few decades later, in the context of the paternalistic ideals promoted by Marshal Rondon, eight reserves were created between 1915 and 1928 (initially with 3,600 ha each, later reduced following land-grabbing by the surrounding farmers) by the newly established, and highly inefficient and corrupt, Indian Protection Service (SPI). This was a process of tacit 'containment' in small reserves, where it was expected that indigenous peoples would be assimilated into the rest of national society (Brand 1997), similar to the policies implemented in the United States at the end of the nineteenth century. Because of the limited space and the agglomeration of different groups and families in very small areas, the reserves have been hotspots for tensions and growing frustration (CIMI 2001). The concentration of people facilitated the utilisation of badly paid indigenous labour, which guaranteed the economic insertion of the Kaiowa in the economy of the frontier and to some extent preserved their (highly subordinate) participation in regional society.

After 1920, large farms began to be opened by migrants attracted to the region from the south of Brazil and incentivised by the national government. A sequence of policies and legislation promoted the privatisation of ecosystems, deforestation and monoculture-based agribusiness. The region where most Kaiowa lives in Southern Mato Grosso is famous for its excellent, red soils—it is claimed by the Kaiowa-Guarani that it is red because it contains their own blood (see Morais 2017)—which were appropriated by the new farmers. Several migratory waves of people from Rio Grande do Sul (known as *gaúchos*) moved to Mato Grosso in the first decades and then again in the 1950s and 1960s following the cancellation of the territorial concession granted to Matte Larangeira and growing competition with *erva-mate* produced in Argentina (leading to the eventual closure of the Larangeira corporation). The Kaiowa were then coerced to leave their traditional lands and witnessed the destruction of the woodlands that give them their name ('people of the forest').[5] New legislation was introduced by the Brazilian military dictatorship in 1973—the Indian Statute—but this basically maintained the assimilationist approach and did not resolve the growing gap between actually existing reserves and the many sites claimed by the Kaiowa as their legitimate land. A more recent phase began in the 1990s with the intensification of an agribusiness-based economy and the exponential growth of plantation farms. Local changes were a direct reflection of national trends and the power of the agribusiness sector, which included a broad coalition between landowners, conservative politicians, banks, industry and transnational corporations (Ioris 2018b). Agribusiness has been naturalised as above party disputes, as something that is supposedly intrinsically beneficial to the country so that any obstacle, including the rights of indigenous groups, must be removed, at any cost. In this sense,

[5]Forcing people out of their land is called *esparramo* or *debandada* in Portuguese and *sarambipa* in Guarani (Brand 1998).

Fig. 7.2 Agribusiness production in Southern Mato Grosso in areas grabbed from the Kaiowa-Guarani

it has largely hijacked the political debate and forced the approval of highly questionable policies and legislation, as in the case of the new forest code, changes in labour regulation and the attraction of international investment funds (Ioris 2017).[6]

Brazilian agribusiness in areas of agricultural frontier, such as Southern Mato Grosso, has given rise to even higher levels of speculation, dispossession of common land and wide-ranging aggression against those with different interests (Fig. 7.2). Frontier making creates favourable conditions for the arrival of unscrupulous individuals in search of rapid enrichment and for the collective acceptance of morally questionable economic and political practices. The recipe for serious socio-spatial conflicts is then complete: on one side, adventurers and speculators reinvented as 'agri-food producers' (the euphemism often used by agribusiness farmers to describe themselves) and, on the other, marginalised native peoples who have been living in the region for many generations in a context of contrasting relations with land and society. The failure to meet indigenous peoples' legitimate demands and the invalidation of even their most basic human rights are clear signs of the institutional racism

[6]The 'bonfire' and destruction of the National Museum of Rio de Janeiro on 2 September 2018 offered a sinister metaphor for the low quality of Brazilian democracy and the neglect of public culture and social justice. Likewise, the pervasive violence against indigenous groups was graphically reproduced on 21 April 2000, on the very site where the Portuguese arrived in the country, during the commemoration of five centuries of Brazilian history.

that pervades the Brazilian state and of how its ideology of modernisation has been constructed at the expense of, and against, subaltern sectors of society. Instead of legal rights over their land, the Kaiowa-Guarani have only been offered backdoor access to regional space, bearing in mind that since the early twentieth century farms have been opened in areas illegally expropriated by farmers and the government. The recurrent violence against the legitimate proprietors of the land—the indigenous groups—is in direct breach of the fundamental principles of the Brazilian constitution. However, the enforcement of the legislation by local judges, civil servants and politicians typically rules in favour of the most powerful economic sectors. This corresponds to the wider problem of non-consensual expropriation of indigenous lands around the world (Doyle 2015).

The dominant ideology of an 'Indian without time' (i.e. idealised, according to the colonial mindset criticised above, as a savage who, to be genuine, has to be confined to the past that permeates school textbooks and the public imaginary) gives support to the idea of an 'Indian without space' (no rights over land, no need for land, no space) because they have lost their right to land insofar as they are no longer seen as 'real Indians'. This is an example of the 'tyranny of authenticity' that threatens to exclude individuals and groups who do not fit some narrow criteria set for membership (Lennox and Short 2016). It has been a regular, everyday experience of social, ecological, cultural and physical violence and a difficult struggle for the Kaiowa to see their rights recognised. During the first visit of Pope John Paul II to Brazil, the Guarani leader Marçal de Souza Tupã-Y denounced the long-standing practice of expropriating land and massacring indigenous peoples, demonstrated by Fig. 7.3. After several unsuccessful attempts by farmers to bribe and quieten Marçal, he was murdered on 25 November 1983, three years after meeting the pope. Nobody, as has so often been the case, was held responsible or punished for Marçal's murder. In fact, public agencies are often driven to provide some kind of response only after a large-scale tragedy is reported by the international media. The bad news kept pouring in: 'High murder rates blight Brazil's indigenous communities' (BBC, 28 February 2014); 'Brazil indigenous leader's killing raises tension' (BBC, 3 September 2015); 'Dispute turns deadly as indigenous Brazilians try to "retake" ancestral land' (The Guardian, 14 July 2016); 'The Guaraní Kaiowà people could soon be wiped out' (Lifegate, 20 April 2017); 'Au Brésil, une manifestation d'Indiens tourne à l'affrontement avec la police' (Le Monde, 26 April 2017). The New York Times reported on 29 May 2017 that in 'August 2015, some of the Guarani-Kaiowa decided to reoccupy part of their territory. They camped on land owned by ranchers. The landholders had other plans: According to reports, they hired armed militias to try to drive the tribe out. Semião Vilhalva, a tribal leader, was shot and killed. There were accounts of torture, rape and child abduction.'

Another famous victim was Ambrósio Vilhalba, a leader of the Guyra Roka ('Place of the Bird') community, who had spent decades campaigning against the planting of sugar cane on the lands of his tribe. He became internationally famous after taking part, as the main character, in the award-winning film *Birdwatchers* (sadly and prophetically, the character was based on himself and, as in the movie, he also ended up being murdered). He travelled the world to speak about the Brazilian

Fig. 7.3 Graves of Indigenous people killed by farmers in Caarapó

government's failure to protect native Guarani land. After months of death threats, in 2013 Ambrósio was found dead in his hut from multiple stab wounds. According to Guarani religious beliefs, death by murder or suicide is not the end of the story but brings additional troubles to all involved. The Kaiowa feel particularly disheartened when a relative is murdered and the body simply disappears, which has unfortunately happened all too often (Morais 2017). Violent deaths without a body to bury are considered to be very dangerous, given that a person has two souls and the bad one [*angue*] will remain like a phantom, threatening the living. Amidst a hyper-violent situation that affects both the dead and the living—further aggravated by the election in October 2018 of a right-wing president, which immediately triggered new waves of cowardly attacks on indigenous groups and many other racial minorities (Fuhrmann 2018)—the Kaiowa-Guarani have reacted according to their means and formed some limited but important alliances with national and international organisations, universities and churches. They have, therefore, disrupted the pace and the configuration of frontier making, leading to a mosaic of spaces, as discussed next.

7.4 … But an Insurgent Indigenous Geography

The Kaiowa-Guarani, among numerous other indigenous peoples in South America, had their territories invaded and sociocultural practices disrupted as part of the wider process of frontier making. The socio-spatial frontier enthralled them like an avalanche and triggered a long process of aggression, intolerance and forced assimilation. The 'Kaiowa problem'—more accurately, the problem brought to the Kaiowa and their forced conversion into a problem, following Ribeiro (2015)—has been an acute process of commodification (in the nineteenth century of forests and labour, and in the twentieth century of land, labour and water), facilitated by the production of racialised spaces that marginalised indigenous populations in their own territory. While the Kaiowa were historically at the margins of official national history, frontier making left them even more ostracised and oppressed than during the colonial years. Out of the four million hectares originally occupied by the Kaiowa in the region, indigenous families were left with around 40,000 ha of regularised land spread across various reserves and resettlements (Benites 2014; Brand 1998; Cavalcante 2014). Some reserves are preposterously small (only a few hectares per family, or even less, as in the case of the Dourados reserve, where 15,000 people live on only 3,000 ha) and land tenure is not clear because of objections from farmers and other business sectors (even after the lengthy process of regularisation). Despite all this, the Kaiowa has demonstrated an unfailing ability to adapt to the new world order and to retain crucial elements of their identity, culture and knowledge that certainly help them in their struggle for land and recognition. Combining multiple strategies, which have included settling in remote corners of the large farms, seeking employment outside the reserves, making their voices heard in public and forming strong internal and external networks, the Kaiowa have managed to engage with the advancing frontier and, in recent years, secured some modest but nonetheless tangible political and territorial successes.

Through the mobilisation of their customs and religious values, the Kaiowa-Guarani have vividly replicated, according to their own circumstances and way of life [*ava reko*], shared elements of the social and agrarian transformation that has been taking place around the world in the last few centuries of capitalist history.[7] It is a dynamic geography of general similarities or commonalities, but also with distinctive cultural and politico-ecological characteristics. Kaiowa's existence is particularly characterised by a scalar and deeply religious conception of space,

[7]The eighteenth century English experience is a case in point described by Thompson (1993), where enclosure took the commons away and made people strangers in their own land. At the same time, the appropriation of common land and reduction of food security in England triggered riots, which were the rational response of poor people with some remaining power to help themselves.

from the household to the network of settlements and indigenous reserves. Following the Kaiowa cosmovision, people do not own or trade the land, but live there, sharing it with other creatures and constantly having to negotiate their conditions; these creatures hold or belong to the spirits [*jaras*], whose categorisation depends on the relations established between humans and non-humans (Pereira 2010). These religious, cultural and social experiences are constantly mobilised in favour of their daily survival and, in cases of dispossession, the prospect of returning to the lost areas. The skilled resistance of indigenous groups demonstrates both the inadequacies of the modernising project and the resilience of non-Western societies when protecting livelihoods and cultural practices on their land. In this way, Kaiowa groups have become one the most active geographical protagonists in the region, constantly trying to adapt and respond to pressures, as well as protect and restore elements of their previous life conditions disrupted by frontier making. This is materialised in the diversity of spaces through which the Kaiowa subvert and complicate the course of frontier making, as described below

Indigenous reserves—despite serious management problems, reserves constitute the most stable spaces available to the Kaiowa, largely accepted by wider society and reasonably well protected by the state. The reserves were established in two main phases. The first involved the creation of the eight SPI reserves, with a total of around 18,000 ha, in the first decades of the twentieth century. The process imposed the 'peasant model' on the Kaiowa, fixing them in certain state-owned areas where the land can be cultivated by the community (Oliveira 1983). This was certainly not a stress-free process, given that not all the reserves were established in the original areas where indigenous families used to live, but merely observed bureaucratic convenience (most of these first reserves even had their area reduced due to pressure from neighbouring farmers and with the help of corrupt civil servants). Against the will and the traditions of the indigenous people, the land in the reserves was divided into private lots, which is directly in conflict with their tradition of common land [*teko-hakuaaha*]; see Figs. 7.4 and 7.5. One very serious issue was the accommodation of different extended families or ethnic groups (Kaiowa-Guarani, Nhandeva-Guarani and Terena) in the same reserve, which only generates new tensions and nurtures disputes. Levels of violence and crime in the reserves are directly or indirectly related to growing frustration with persistent abandonment, racism and discrimination. Stories about youth violence in the reserves are favourites in local and national media, particularly those involving sexual assaults, gang killings and drug trafficking.

The second phase of reserve formation was launched by the approval of a new Brazilian constitution in 1988, which officially recognised indigenous lands (Article 231) as areas traditionally and permanently occupied and indispensable for indigenous peoples' productive activities, for preservation of environmental resources needed for their well-being and for physical and cultural reproduction in terms of uses, customs and traditions. The same constitutional article determines the nullification of any act that led to the occupation or appropriation of indigenous lands, stating that these are inalienable and the rights over them are imprescriptible. The approval of the new constitution coincided with the adoption of the Indigenous and Tribal Peoples Convention (ILO-Convention 169) in 1989 and the growing international importance

Fig. 7.4 Indigenous family household in the Pirajuí reserve

of the principle of 'free prior and informed consent' (FPIC). The problem for the Kaiowa-Guarani has been the lack of enforcement of these formal rights and their refusal to consent to the confiscation of their lands. There is always a protracted struggle through the various layers of the judiciary, with endless appeals and explicit political pressure exerted on local judges (normally themselves large landowners and members of wealthy regional families). Among the 51 existing reserves, only 24 are reasonably stable, which does not mean they are free from tensions (Zelic 2018). Policy failures and lingering demands are rationalised by politicians and authorities on the grounds of the need to follow due process, but really to benefit the agribusiness sector and because of the social need to accept the costs of development (as is often the case, victims are blamed for their own misfortune).

Roadside encampments—Because of organisational, behavioural and demographic problems in the official reserves, a significant proportion of the Kaiowa-Guarani people (around a quarter of the total population) have opted to move out and live in encampments next to the main motorways and secondary roads. These encampments can either be relatively permanent sites of residence (there are cases of families who stay in such dreadful conditions for many decades) or temporary campsites for families hoping to move back to reclaimed areas [*retomada*]. In both cases, people living in encampments maintain close connections with relatives in other sites and in the reserves, always with the expectation of better conditions and the hope to return to the areas where they or their families used to live. The Kaiowa

Fig. 7.5 A Kaiwoa-Gauarani with the Author, before a football match in the Pirajuí Reserve

have resolutely resisted the antagonistic rationality that led to the fragmentation and privatisation of space, closely promoted and coordinated by the national state (Barbosa and Mura 2011). The space of the encampment, regardless of the unpleasant and insalubrious circumstances, is also a space of anticipation and potentiality. This ambivalent situation, among others, is presented in the documentary *Martírio* by Vincent Carelli, released in 2017, which shows how precarious and risky conditions are for the Kaiowa, but also their firm commitment to survive and honour their ancestors. There is a constant threat posed by farmers and paramilitaries (as aforementioned, militias hired by farmers and rural companies, normally employing retired or active policemen). There are also serious dangers of road traffic accidents and fire coming from sugar cane fields ahead of the harvest. Despite recurrent cases of atrocious violence, the police express little interest in finding those responsible for the crimes, and the judicial system is quite unprepared to punish. These forms of treatment only fuel resentment and, paradoxically, lead more individuals to join roadside encampments ahead of future *retomadas*.

Areas of *Retomada*—literally, *retomada* means ‘taking back’, but more significantly it implies reaction and spatial reoccupation (social movements in Brazil reject the politically derogatory term ‘invasion’). The *retomada* is the return to the original places where the older Kaiowa generations were expelled in the 1950s or 1960s (many elders are still alive and able to testify to their connection with the intended places). Through the *retomadas*, ‘indigenous families reoccupy areas where they can carry out their community life, establishing their dwellings, planting smallholdings, and practicing their ritual and religious life’ (Oliveira 2018: 12). The *retomada* is a political expedient of peasant-like resistance that relies on the construction of material, organisational and ideological means for its utilisation (Ferreira 2007). In Guarani, *retomada* is described as *jeike jey* (*jeike* is reflexive prefix insofar as that means ‘we enter’ or ‘let’s enter’ when used with an inclusive verb form, and *jey* means again; therefore, *jeike jey* literally signifies ‘we enter again’, which can be translated as return or *retomada*). After a frustrating wait and exasperation with unfulfilled promises, indigenous political and religious leaders realised that the official approach of reserve creation was misleading and would never address their economic and cultural needs. These leaders (invariably, obvious targets for new assassinations) organised a group to take action and evict the actual invaders (the current farmers) and make an attempt to self-demarcate their legitimate lands. *Retomadas* are moments of rupture when dominant politics is subverted and indigeneity can emerge through ‘an insurgence of indigenous forces and practices with the capacity to significantly disrupt prevalent political formations’ that render ‘illegitimate the exclusion of indigenous practices from nation-state institutions’ (de la Cadena 2010: 336).

The fundamental difference between the original SPI reserves and *retomada* areas is that the operationalisation and risks involved in these initiatives are exclusively those of the indigenous groups, beyond the tutelage of the state. See Figs. 7.6, 7.7 and 7.8. ‘Given the nonexistence of other efficient alternatives, the retakings [*retomadas*] have turned into the main strategy of indigenous people for recognition of their territorial rights, at present, having been incorporated as a flag of struggle by the indigenous movement…. They constitute a post-tutelary form of the exercise of the policy by the Indians, implying a different mode of conceiving their relationship with the State’ (Oliveira 2018: 13). CIMI registered at least 88 areas demanded by the Kaiowa-Guarani, but the list is certainly much longer (Morais 2017) and the struggle is now to recover plots much larger than the old reserves (*retomadas* from farmers with more than 10,000 ha), which of course infuriates the agribusiness community and allied judges and politicians. It is crucial to note that this autonomous movement is more than just a form of indigenous agrarian reform, but expands into an intense site of spatiogenesis where traditions, new influences and articulation with other sites and other groups converge to consolidate the newly retaken land. Ethnographic work has revealed that the *retomadas* ‘imply deep movements of cultural revitalisation and social and political reconfiguration for these peoples’ (Oliveira 2018: 14).

Globalised Kaiowa space—another important space competently produced by the Kaiowa as part of their political agenda of survival and territorial affirmation

Fig. 7.6 *Retomada* in the vicinity of the Dourados reserve

Fig. 7.7 Private security GUARDS (paramilitaries) hired by farmers to contain a *Retomada*

Fig. 7.8 Indigenous family in a *Retomada* in a Soybean Farm in Caarapó

is an international arena of protest, involving a network that includes academics, artists and multilateral organisations. In the last few years, various Kaiowa leaders have been interviewed by global media channels, invited to speak in international forums and taken part in publications, movies and documentaries. The insertion of the Kaiowa in the globalised space is evidently related to broader indigenous movements around the world (see Lennox and Short 2016). All this has attracted growing attention from the international community, meaning that the movement can be compared with the prominent campaign articulated by the Zapatistas (although of course the Kaiowa do not have anything resembling an army of national liberation). The activism of the Kaiowa in national and international circles indicates that their leaders have realised the importance of employing an effective, articulate discourse and learning to engage with non-indigenous players. Public outrage over the Kaiowa genocide has soared across the world and triggered multiple reactions.

On 24 November 2016, the European Parliament approved a resolution that strongly condemned 'the violence perpetrated against the indigenous communities of Brazil', deplored 'the poverty and human rights situation of the Guarani-Kaiowa population in Southern Mato Grosso', reminded 'the Brazilian authorities of their obligation to observe international human rights standards with respect to indigenous peoples' and, among other things, expressed 'concern about the proposed constitutional amendment 215/2000 (PEC 215), to which Brazilian indigenous peoples are fiercely opposed, given that, if approved, it will threaten indigenous land rights by making it possible for anti-Indian interests related to the agro-business, timber, mining and energy industries to block the new indigenous territories from being recognised.'[8] PEC 215 is essentially a canny manoeuvre by the Brazilian parliament to stop the creation of new indigenous reserves (which would become an exclusive prerogative of the agribusiness-dominated congress).[9] This would have terrible consequences for most Kaiowa who have been forced from their ancestral lands.

Spectral space—the Kaiowa-Guarani cultural background contrasts with the rigid, artificial limits of private property institutions, which constrain social mobility and interfere with traditional agricultural practices based on the rotation of cultivation sites. According to Kaiowa-Guarani traditions, the space required for a family needs to be constantly regenerated and reconfigured through movement in the landscape that is essential for the fulfilment of their way of life. Because of this, the consequences of land commodification and the advance of the agribusiness frontier were extremely negative for these indigenous groups. Even so, in a way that was unanticipated by governments and farmers, the Kaiowa have resisted and recaptured some tracts of land, and their population is on the rise. They have represented an uncomfortable presence in the regional economy and a nuisance from the perspective of local elites; a threat constantly present in the Guarani names for places and things and the alarming presence of 'these strange people' in roadside encampments who, from time to time, come together to demand another area they associate with their heritage. Their leaders have repeatedly emphasised that their long-term, non-negotiable goal is to reinstate most of the original Kaiowa territory instead of the small islands so far granted to them by the state (Barbosa and Mura 2011). The political crux of the matter is that the land demanded by the Kaiowa is legitimately theirs not only because indigenous populations used to live in these areas (which is the main legal stipulation for the return of ancestral land), but more importantly because the possibility of a meaningful future for them fundamentally depends on their physical and metaphysical interdependencies with this land. In other words, the Kaiowa, just like other subaltern and proletarian groups, are politically relevant not because of their past, but because of their future based on anticipated ownership of the land (as Fig. 7.9 shows, the Kaiowa reject the current spatial configuration and use road signs to affirm ownership and the right to name their own land).

[8]Those worries attracted another resolution by the European Parliament, approved on 3 July 2018, against the violation of the rights of indigenous peoples, including the impacts of land-grabbing.

[9]Another extremely controversial measure, adopted by the Supreme Federal Court and now under analysis in the same parliament, is the '*marco temporal*' (arbitrary cut-off date for land claims).

Fig. 7.9 Indigenous name of the land written over farm signposts

Kaiowa geography evolves in different directions and contains multitemporal elements from their past that shape the visible present and the desired future. A crucial part of this geography is the large 'space in waiting' that is dreamed about, constantly narrated and re-narrated for younger generations, an anticipation of a concrete future reality that will reinstate the past. It constitutes the spectre of a lost world that nonetheless will be restored one day if they continue to actively envisage it. The distribution of the existing reserves on the map shows this palpable spectre (the vast land in-between reserves and encampments), a constant reminder of the perennial defiance and forthcoming action of the Kaiowa to recover land that is theirs.[10] Their spectral condition is a shameful sign of the misplacement of things and, to the frustration of politicians and farmers, they have national law and universal human rights in their favour, although still poorly enforced. They have the moral high ground of those who were brutally invaded and subjugated for spurious material gain, similar to the Pashtuns of Afghanistan, who can always afford to expend time and wisdom against invading superpowers (Britain, Russia and the USA), which have displayed great incompetence in their dealings with more than Western mindsets that operate beyond the language of money and bullets.

The intention is to restrict legal rights to claim traditional territories to land physically occupied on 5 October 1988 (the date of approval of the current constitution).

[10]It can be consulted at https://terrasindigenas.org.br.

There is nowhere the Kaiowa spectre causes more anxiety and theatrical vociferation than in the Brazilian House of Representatives, notoriously dominated by agribusiness interests. Congressmen of the Parliamentary Farming and Cattle Raising Front [*Frente Parlamentar da Agropecuária*], in particular like to put on a show in front of the cameras to impress voters and supporters. The 'Indian problem' is a juicy topic, able to mobilise parliamentarians from different parts of the country, and in this context the Kaiowa-Guarani represent the most combative threat and a 'bad example'. For instance, in a public audience of the Agricultural Commission held on 08 May 2013, with the presence of several Secretaries of State, to discuss growing demands for the rehabilitation of indigenous areas (the main focus was on Southern Mato Grosso), some of the more vocal congressmen spoke as follows:

> Is the Government's goal for 25% of Brazilian territory to be indigenous areas? I have seen a FUNAI[11] report saying that the aim is for 25% of Brazilian territory to be indigenous areas. It says that there are 611 areas due to be created as indigenous lands… A new indigenous Guarani nation is about to be born. (Valdir Colatto)

> There was no solution, and the problems were getting worse… We are seeing, in this process, orchestration, abuses, the pressure that the Indians make… [The Indians] leave their land to create other reserves… If we do not take action, they will be creating Yanomami, Guarani, and other nations. And they have the support of the King of Norway, who funds NGOs… I am of German origin - my ancestors came from Germany in 1853 - but I am Brazilian, I defend this country and I will defend it intransigently. We cannot give into this pressure… (Luis Heinze)

> In Southern Mato Grosso, we have had numerous invasions of rural properties, with the support of FUNAI, to where Indians are taken, clearly with documents from Paraguay. (Reinaldo Azambuja)

Leaving aside the poor syntax of the strident politicians, the acrimonious, racist and bitter speech suggests that the establishment is notably on the defensive and at least deeply disturbed by the indigenous peoples' moral and legal upper hand. Hegemonic groups and their political representatives are less in control than they would like to be. This meeting of the Agricultural Commission was organised a few days after the flamboyant invasion, in the middle of a plenary session, by dozens of indigenous leaders (17 April 2013) in protest against PEC 215 (see above). Dialogue seems quite impossible (as Marx put it, 'force is the arbiter'), even more so in a National Congress where a significant proportion of members (who also own large farms and are often involved in massive corruption scandals) sponsor a clear anti-indigenous agenda. Numerous bills have been introduced aiming to erode legislation on the regularisation of indigenous areas, and deepening the commodification of

[11]FUNAI (the National Indian Foundation) is the government body that deals with policies relating to indigenous peoples. In 2019, it was removed from the Ministry of Justice and became part of the Ministry of Human Rights, while responsibility for the demarcation of indigenous areas was transferred to the Ministry of Agriculture (controlled by the agribusiness sector and openly against the recognition of indigenous land rights). The minister appointed by the new president in January 2019, Ms. Tereza Cristina, is a large-scale farmer and a congresswoman representing Southern Mato Grosso, who selected as close advisors a team of lawyers who had fiercely attacked the Kaiowa-Guarani on behalf of landowners.

labour and nature has been a top priority in recent times in Brasília (CIMI 2018b). This reinforces the paradoxical situation that those who are ostensibly vulnerable and powerless are clearly able to alarm those who are nominally powerful, but who have fundamentally abused their mandate and turned their back on justice and democracy.

7.5 Reconstruction of the Kaiowa Space

Morirò… ma invendicato
il mio nome non sarà,
il mio sangue avvelenato
mille morti costerà!

Pery, in the opera *Il Guarany*, by Carlos Gomes (1870)

The Kaiowa, despite their troubles and the abject violence they have experienced, have written one of the most intriguing stories of contemporary, multicultural Brazilian society and its perennial dependence on frontier activity. It is a situation fraught with tensions and hidden complexities. From the perspective of indigenous groups, frontier making is a process of hyperbolical action that generates its own antipode, as movement in one direction often ultimately leads to opposite results.[12] The Kaiowa-Guarani have had to resist an antagonistic rationality that invariably led to the fragmentation and privatisation of space, promoted and coordinated by the national state. There has been a terrible banalisation of aggression, regular assassinations and most of their ancestral land is still a spectre, which seems to prove that the Kaiowa are merely on the losing side of regional development. Their land was grabbed, their social life violently disrupted, their world will never be the same. However, the experience of frontier making endured by the Kaiowa demonstrates that indigenous groups are, in effect, both victims and protagonists. The very presence of indigenous peoples in the region, after decades of abuse, reveals a remarkable capacity to cope, despite all the difficulties, with the negative impacts of the advancing frontier and suggests that such groups are among the most resilient and skilled of those embroiled in frontier making. It is true that the impacts of colonisation affected the Guarani groups later than the majority of other indigenous peoples (for a number of historical reasons, including the Jesuit missions and their hiding in the forests), but it is also the case that their recent history has been marked by extraordinary courage in handling market-based globalisation, the commodification of common resources and the homogenisation of culture.

[12] In the words of Galeano (1982: 63), '*Colón, buscando el Levante, ha encontrado el Poniente. Leonardo adivina que el mundo ha crecido.*' [Columbus, looking for the Levant, has found the West. Leonardo guesses that the world has grown.]

Indigenous peoples have commonly been portrayed in mainstream Brazilian political debate using normative language which represents them as responsible for their own condition. In this way, their action has been judged against criteria and values that are foreign to them. This is an *ex post facto* construction of their fate that leads to further misunderstandings. Many of the subtle elements in their culture and value system are not easy to bring into the realm of westernised academic comprehension, but it is not too difficult to perceive a wealth of sensibilities that help to fill many of the social and ethical gaps of frontier making. The Kaiowa have a fundamentally different association with the landscape where their ancestors were buried and still are, given that for them both the living and the dead reside in that land; their life, identity and existence depend on that land, and at the end of their life they become land themselves. Areas beyond the farm fences are remembered, evoked, celebrated and, when the time is ripe, claimed and ultimately recovered. What is regularly missed in conventional accounts is the truly latent geographical agency of the Kaiowa, that is, their ability and willingness to remember and dream of a different spatial setting, organised according to their culture, values and knowledge. After the humanist lessons of Yi-Fu-Tuan, we know that geographical agency not only includes physical action, but also has a crucial subjective and symbolic dimension. The geographical agency of the Kaiowa—which is basically an expression of a political ontology—is translated into a landscape of potentialities in which the indigenous groups struggle to survive in the interstices of agribusiness modernisation and constantly nurture their preparations for a different spatial reality erected on the basis of a lived cultural heritage.

The recognition of such latent agency unveils the tripartite ontological arrangement described by Deleuze (1968) at the intersection of the interdependent 'registers': virtual, intense and actual. According to Deleuze, intensive morphogenetic processes follow virtual multiplicities to produce localised, actual realities with extensive properties. Unlike how it is normally described, the political power of the Kaoiwa is maintained independently of their immediate control of the territory, but it is accumulated as latent geographic agency that is eventually manifested in the *retomadas*. The virtual is actualised by way of intensive processes. Because of their geographical agency, the Kaiowa-Guarani have managed not only to recuperate their religious traditions, but to mobilise these in support of their social identity, political voice and spatial strategies. Their search for a better life is informed by the symbolism of a mythic land of peace and plenty; different Guarani groups have distinct myths and narratives of the 'land without evil' but for the Kaiowa the myth is certainly a source of hope and helps them in the difficult journey back to the lost areas (Chamorro 2010). Notwithstanding all the profound changes and influences that have affected the Guarani, their most cherished religious and existential creeds continue to underpin social values and interpersonal relations. According to Clastres (1975), the search for the new world has carried on for centuries through physical movement in the South American territory, led by powerful spiritual leaders to overcome existing circumstances and sociopolitical crises. In this sense, the religiosity that infuses Guarani cosmology operates as a refuge and has left them prepared to cope with contemporary economic frontiers. Against all the odds, the Kaiowa-Guarani have revealed deep persistence and wisdom in using their ancient religious beliefs to help them deal with twenty-first century challenges.

The theological thought of the Kaiowa corresponds to a particular time-space conceptualisation of the world, where deities enter households and establish relationships (Mura 2006). The main locus of geographical agency is the interface between the extended family (with around 100 individuals) and the larger indigenous network that connects different reserves and settlements. The basic spatial unit is the *tekoha*, the specific area where one or a few extended families live, strongly connected with other *tekohas* through regular meetings, marriages and ceremonies. A *tekoha* has no rigid limits, but comprises the space needed to hunt and fish, is broadly delimited by hills and rivers, and often includes a river basin (Benites 2014). Its meaning-content has more to do with the production of culture than with economic production (Meliá 1989, in Brand 1998: 23). According to the Kaiowa-Guarani tradition, when there is a disagreement or need for more resources, part of the group moves to another area and establishes a new *tekoha* (Silva 2007). Therefore, spatial mobility [*guata*, which means wandering around, walking or perambulation] is an important pillar of Kaiowa-Guarani culture and maintaining a good way of life (Brand 1998). The movement associated with *guata* suggests a constant willingness to be free to resettle elsewhere in order to maintain the livelihood of the family and to guarantee social reproduction. Through *guata* there is also contact with the *tekohas* lost to economic development and a perennial longing for return to such areas that still show marks of indigenous activity decades ago (burial grounds, old settlements, cultivation plots, etc.).

According to Kaiowa religiosity, these lost areas remain populated by gods and invisible entities, some benign, some evil, so people have to prepare themselves spiritually before they can return. The radicalism of indigenous spatial action (incomprehensible for the rest of society and apparently illogical, considering the level of killing and suffering they have experienced) is guided and anchored by the invisible world of their ancestors and the dream of a land that will restore the desired connections between the gods, the dead and the present generations. The latent geographical agency—a true geography of potentialities—is directly and powerfully fuelled by the extraordinary emphasis on eschatology in the Tupi-Guarani world (which includes the Guarani nations), that is, their social life may be relatively simple but the taxonomy of the supernatural world is complex and this spiritual dimension has an active, intense presence in quotidian, material life (Viveiros de Castro 1986). Several scholars who have studied the Guarani have recorded the pre-eminence of religion over all social spheres and, more significantly, the practice of religion as a decisive locus of cultural resistance. The use of their own language, impenetrable to most outsiders, also works as a secret war code that helps to preserve their identity and enforce the meaning of their objects, actions and traditions. New generations are more proud of being indigenous, which is an important legacy of the last four decades of mobilisation and dialogue with allied non-indigenous groups. Clastres (1975) and several other authors have demonstrated the crucial role of religion in the rationalisation of dramatic socio-spatial changes that have affected the Guarani; their religious beliefs are centred around the messianic expectation of the 'land without evil' that will eventually replace the imperfect, mundane reality of the lived world (although this myth has been questioned as a decontextualized academic fabrication, see Villar and Combès 2013). Their cosmology is also informed by the apocalyptic vision of the end of the world, which is now often associated with the inferno of monoculture farms (Morais 2017).

Benites (2014), an indigenous scholar, provides a vivid description of the violence historically experienced by the Kaiowa, but also the sophisticated preparation for *retomadas*, making use of cultural and religious heritages as an integrating and transformative force. The aim of the *retomada* is to restore the lost *tekoha guassu*, which is the territory shared by several extended families and following influential religious and political leaders. It is possible to ascertain from the narrative offered by Benites the crucial political and existential connection between families and their leadership revealed in the dialectics *tekoha – tekoha guassu*. The strategies of land reoccupation are intensely discussed and enacted in the large assemblies [*aty guassu*] organised by the Kaiowa since 1979, in which religious rituals are of paramount importance. Participants identify mutual needs, share tactics, make collective decisions, and prepare documents for public dissemination. The passionate ritualization of their practices and the importance of religion for their political action encourage them to fight. This is a moment of great risk, a real war, but it is more than a holy war against the invaders: it is a necessary struggle to maintain their world. During the *aty guassu* the core assault group is selected, formed from religious people, their assistants, political chiefs, elders and children. This vanguard party spend months preparing themselves for the attack, praying and taking part in strenuous rituals. The four nights before the *retomada* is a time of even more intense religiosity, when the warriors are baptised, which is required in order for them to be recognised and accepted by the dead ancestors and to be protected against evil spirits and invisible beings. The night before, they paint their faces and parts of the body with *urucum* (a plant used to make red body paint) and the males hold their bows and arrows tightly as a sign of respect for the ancestors. After this long preparation, they march for around ten kilometres during the night to collectively retake the land. If everything goes as planned, they immediately build huts and start to fish and hunt to feed the group. A new altar for the continuation of religious ceremonies is also erected. Using tactics like this, in recent years the Kaiowa have managed to recover more than 20 areas, although the farmers who claim ownership of the land have reacted in different ways, frequently through the use of brutal violence (many dozens of indigenous leaders have been murdered because of this open war, including Benites's own father, who died in mysterious circumstances). In the end, this is the dialectics of suffering and healing that profoundly marks the contemporary frontiers of capitalism.

7.6 Conclusions

The previous pages have examined how the straightforward image of modernity and innovation that underpins contemporary development frontiers needs to be replaced with a much more complex picture. Frontier making, which has been a central politico-spatial driving force of capitalist development, remains a highly paradoxical phenomenon, in which progress and abundance are repeatedly promised, while the reality on the ground continues to be shaped by the old practices of exploitation, exclusion and racism. The landscape of the frontier seems simultaneously logical, organised and chaotic, out of place. In the case of Southern Mato Grosso, agribusiness appears novel, but in fact it recreates elements of the colonial past, particularly in the context of violence against local indigenous groups. The Kaiowa-Guarani, among other peoples, are commonly depicted as living examples of stone-age savages, although their geographical practices demonstrate a sophisticated ability to comprehend and creatively react to socio-spatial pressures and economic changes. It is a situation fraught with puzzles and ambiguities; to a large extent, it is the Kaiowa who are offering innovation, while agribusiness encompasses inbuilt obsolescence. Agribusiness seems new, but it mobilises and is justified through practices introduced in colonial times, while indigenous people are historically old, but their reactions, creativity and aspirations are closely connected with contemporary debates on alternatives to development, market globalisation and cultural homogenisation. All this is happening in a highly politicised landscape where indigenous groups, despite all the tragedy, suffering, humiliation and severe neglect by the state, are in effect securing small, but precious, territorial victories. To the surprise of some urban and business groups, the Kaiowa have shown latent geographical agency shaped by religious practices, strong family ties and the ability to internally negotiate the return to their original areas. The wisdom and resistance of Kaiowa-Guarani groups derive from the simultaneous ethnicisation of space and spatialisation of culture. Far from any sentimental romanticism, we can learn that for the last forty years the Kaiowa have been able to regain confidence, mobilise their language, culture and religion, and form strong networks between families and localities to both resist the trend of violence and, when opportunities arise, retake their long lost land.

Different indigenous groups will have diverse levels of association with the Westernised model of economic development (for an overview of development biases, see Escobar 2012), but the manifestation of indigenous culture represents a challenge to the prospects of frontier making and reveals its ingrained contradictions in terms of socioecological violence, social exclusion and inequalities. Contrasting with the narrow rationality of agribusiness farmers and their political allies, the cosmovision of the Kaiowa encapsulates multiple layers in which the material and spiritual terrains converge in a way that allows them not only to labour in the areas currently occupied, but also to almost touch the spectral space that will be returned to them one day. The land of their ancestors belongs to the living descendants and the return to those areas depends, fundamentally, on the initiative and courage of present generations. Because Kaiowa land has a purely qualitative value, which is absolute, perpetual and beyond monetisation, the only logical attitude is to continue the struggle to the last drop of blood. From their perspective, they are witnessing a 'territorial pulse', that is, their lands are only temporarily lost and are there to be reconquered; the Kaiowa never gave up their land; they could not, because it is part of their existence to be returned to the ancestral land. There are many lessons to be learned here, in particular, the talent of the Kaiowa to absorb the increasing and dissimulated brutality of frontier making and, at the same time, voice their political demands, form solid strategic alliances and coordinate land recovery initiatives. It is the case that the affirmation of indigenous identities[13] and the pursuit of long-pending rights are relatively recent phenomena in Brazil and other South American countries, directly associated with the progressive strengthening of democratic reforms. In that context, the resistance and agency of indigenous groups, who are increasingly trying to restore valued elements lost to national development, are crucial components of a wider mobilisation for social and environmental justice.

This results of this analysis endorse the growing importance of indigenous geography in the early twenty-first century, a period characterised by sustained attacks on many of the important social and political achievements of the last two centuries (such as universal equality, rejection of racism and discrimination, basic human rights, etc.). The territorial and agrarian struggle of the Kaiowa-Guarani constitutes an emblematic chapter of a geographical mobilisation in the Global South of the planet, which challenges the conventional, Westernised narrative of modernity or postmodernity (Ioris 2018c). The survival and expansion of groups like the Kaiowa actually represent an 'inconvenient' reminder that other worlds are possible and, quite conceivably, necessary. As social, ethical and political tensions increase, perhaps Brazilians can find helpful responses from those traditionally ignored: the descendants of the early inhabitants of the continent, who may hold answers to some of the problems accumulated through a highly uneven process of national development. Indigeneity is a relational construct and, because of the incomplete erasure of colonisation, indigenous and non-indigenous identities co-constitute each other (Coombes et al. 2011). The Kaiowa, other Guarani populations and the more than 300 first nation peoples have a lot to offer in the collective search for a more meaningful way of life. Perhaps a biblical analogy can even be invoked in this context: non-indigenous Brazilians are the prodigal sons and the indigenous the wise fathers (Fig. 7.10). A huge dose of humility and willingness to learn will certainly be needed in order to move beyond the tragic geography of frontier making.

[13]A new, hybrid type of suburban culture (for example, through hip-hop and similar musical rhythms) is increasingly present, particularly in the reserve near Dourados, and is starting to be followed by other young Brazilians, who combine indigenous symbolism with other sources of artistic inspiration and are skilled users of internet-based communication technologies (de Mari 2019).

Fig. 7.10 Indigenous woman in a *Retomada*

References

Anthias P (2017) Ch'ixi landscapes: indigeneity and capitalism in the Bolivian Chaco. Geoforum 82:268–275

Barbosa PA, Mura F (2011) Construindo e Reconstruindo Territórios Guarani: Dinâmica Territorial na Fronteira entre Brasil e Paraguai (Séc. XIX-XX). Journal de la Société des Américanistes 97–2:287–318

Benites T (2014) Recuperação dos Territórios Tradicionais Guarani-Kaiowá: Crónica das Táticas e Estratégias. Journal de la Société des Américanistes 100–2:229–240

Brand A (1997) O Impacto da Perda da Terra Sobre a Tradição Kaiowá/Guarani: Os Difíceis Caminhos da Palavra. PhD thesis, PUCRS, Porto Alegre

Brand A (1998) "Quando Chegou esses que são Nossos Contrários": A Ocupação Espacial e o Processo de Confinamento dos Kaiowá/Guarani no Mato Grosso do Sul. Multitemas 12:21–51

Cavalcante TLV (2014) Demarcação de terras indígenas kaiowá e guarani em mato grosso do sul: Histórico, desafios e perspectivas. Front: Rev História 16: 48–69

Chamorro G (1998) A Espiritualidade Guarani: Uma Teologia Ameríndia da Palavra. São Leopoldo, Sinodal
Chamorro G (2010) Imagens Espaciais Utópicas: Símbolos de Liberdade e Desterro nos Povos Guarani. Indiana 27:79–107
CIMI (2001) Conflitos de Direitos sobre as Terras Guarani Kaiowa no Estado do Mato Grosso do Sul. São Paulo, Conselho Indigenista Missionário
CIMI (2018a) Violência contra os Povos Indígenas no Brasil: Dados de 2017. Brasília, Conselho Indigenista Missionário
CIMI (2018b) Congresso Anti-Indígena: Os Parlamentares que mais Atuaram Contra os Direitos dos Povos Indígenas. https://cimi.org.br/wp-content/uploads/2018/09/congresso-anti-indigena.pdf
Clastres P (1974) Le Grand Parler: Mythes et Chants Sacrés des Indiens Guarani. Editions du Seuil, Paris
Clastres H (1975) La Terre Sans Mal: Le Prophétisme Tupi-Guarani. Editions du Seuil, Paris
Coombes B, Gombay N, Johnson JT, Shaw W (2011) The challenges of and from Indigenous geographies. In: Del Casino VJ, Thomas M, Cloke P, Panelli R (eds) The companion to social geography. Blackwell, Oxford, pp 523–542
Coombes B, Johnson JT, Howitt R (2014) Indigenous geographies III: methodological innovation and the unsettling of participatory research. Prog Hum Geogr 38(6):845–854
de la Cadena M (2010) Indigenous cosmopolitics in the Andes: conceptual reflections beyond "politics". Cult Anthropol 25(2):334–370
de Mari J (2019) Geração Z Indígena une Apps e Tradição para se Politizar na Era Bolsonaro. https://tab.uol.com.br/noticias/redacao/2019/02/18/geracao-z-indigena-unem-apps-e-tradicao-para-se-politizar-na-era-bolsonaro.htm
Deleuze G (1968) Différence et Répétition. PUF, Paris
Doyle CM (2015) Indigenous peoples, title to territory, rights and resources. Routledge, London
Egoz S, Makhzoumi J, Pungetti G (eds) (2011) The right to landscape: contesting landscape and human rights. Routledge, London and New York
EMGC (Equipe Mapa Guaraní Continental) (2016) Cuaderno del Mapa Guaraní Continental: Pueblos Guaraníes en Argentina, Bolivia, Brasil y Paraguay. Campo Grande, EMGC
Escobar A (2012) Encountering development: the making and unmaking of the third world. Princeton University Press, Princeton and Oxford
Ferreira AC (2007) Tutela e Resistência Indígena: Etnografia e História das Relações de Poder entre os Terena e o Estado Brasileiro. PhD thesis, Museu Nacional/UFRJ, Rio de Janeiro
Fuhrmann L (2018) Ameaças e Ataques a Indígenas e Sem-terra Marcam Fim de Semana da Vitória de Bolsonaro. https://deolhonosruralistas.com.br/2018/10/29/ameacas-e-ataques-a-indigenas-e-sem-terra-marcam-fim-de-semana-da-vitoria-nas-urnas-de-bolsonaro
Galeano E (1982) Memorial del Fuego I: Los Nacimientos. Siglo XXI, Madrid
Ingold T (1993) The temporality of the landscape. World Archaeol 25(2):152–174
Ioris AAR (ed) (2012) Tropical wetland management: the South-American Pantanal and the international experience. Ashgate, Farnham, Surrey
Ioris AAR (2017) Agribusiness and the neoliberal food system in Brazil: frontiers and fissures of agro-neoliberalism. Routledge, London
Ioris AAR (2018a) Amazon's dead ends: frontier making the centre. Polit Geogr 65:98–106
Ioris AAR (2018b) The politics of agribusiness and the business of sustainability. Sustainability 10(5):1648. https://doi.org/10.3390/su10051648
Ioris AAR (2018c) Seeding a narrow future and harvesting an exclusionary past: the contradictions and future scenarios of agro-neoliberalism in Brazil. Futures 95:76–85
Jackson P, Penrose J (eds) (1993) Constructions of race, place and nation. Minnesota University Press, Minneapolis
Larsen SC, Johnson JT (2012) In between worlds: place, experience, and research in Indigenous geography. J Cult Geogr 29(1):1–13
Larsen SC, Johnson JT (2017) Being together in place. University of Minnesota Press, Minneapolis

Lennox C, Short D (eds) (2016) Handbook of Indigenous people's rights. Routledge, London and New York
Löwy M, Sayre R (2018) Raymond Williams, romanticism and nature. Capital Nat Soc 29(2):75–91
Mariátegui JC (2007[1928]) Siete Ensayos de Interpretación de la Realidad Peruana. Fundación Biblioteca Ayacucho, Caracas
Markus C (ed) (2013) Kaiowá: Um Povo que Caminha. São Leopoldo, ISAEC/DAÍ-COMIN
Martins JS (1995) Os Camponeses e a Política no Brasil, 5th edn. Petrópolis, Vozes
MacDonald K (2017) My experiences with Indigenist methodologies. Geogr Res 55(4):369–378
Melià B (2004) El Pueblo Guaraní: Unidad y Fragmentos. Tellus 6:151–162
Morais BM (2017) Do Corpo ao Pó: Crônicas da Territorialidade Kaiowá e Guarani nas Adjacências da Morte. São Paulo, Elefante
Mura F (2006) À Procura do "Bom Viver": Território, Tradição de Conhecimento e Ecologia Doméstica entre os Kaiowa. PhD thesis, Museu Nacional/UFRJ, Rio de Janeiro
Murton B (2012) Being in the place world. J Cult Geogr 29(1):87–104
Ochigame R, Holston J (2016) Filtering dissent: social media and land struggles in Brazil. New Left Rev 99:85–108
Oliveira JP (1983) Terras Indígenas no Brasil: Uma Tentativa de Abordagem Sociológica. Boletim do Museu Nacional – Antropologia, no. 44
Oliveira JP (2018) Fighting for lands and reframing the culture. Vibrant 15(2):1–21
Panelli R (2008) Social geographies: encounters with Indigenous and more-than-White/Anglo geographies. Prog Hum Geogr 32(6):801–811
Parson S, Ray E (2018) Sustainable colonization: Tar Sands as resource colonialism. Capit Nat Soc 29(3):68–86
Pereira LM (2010) Demarcação de Terras Kaiowa e Guarani em MS: Ocupação Tradicional, Reordenamentos Organizacionais e Gestão Territorial. Tellus 18:115–137
Pereira LM (2016) Os Kaiowá em Mato Grosso do Sul: Módulos Organizacionais e Humanização do Espaço Habitado. Dourados, UFGD
Quijano A (2008) El Movimiento Indígena y las Cuestiones Pendientes en América Latina. Cotidiano - Revista de la Realidad Mexicana 151:107–120
Radcliffe SA (2017) Geography and indigeneity I: indigeneity, coloniality and knowledge. Prog Hum Geogr 41(2):220–229
Ramos AR (1994) The hyperreal Indian. Crit Anthropol 14(2):153–171
Ramos AR (2018) Por una Crítica Indígena de la Razón Antropológica. Anales de Antropología 52(1):59–66
Ribeiro D (2015) O Brasil como Problema, 2nd edn. São Paulo, Global
Riddell JK, Salamanca A, Pepler DJ, Cardinal S, McIvor O (2017) Laying the groundwork: a practical guide for ethical research with Indigenous communities. Int Indig Policy J 8(2). https://doi.org/10.18584/iipj.2017.8.2.6
Rogers TJ, Bain S (2016) Genocide and frontier violence in Australia. J Genocide Res 18(1):83–100
Shaw WS, Herman RDK, Dobbs GB (2006) Encountering indigeneity: re-imagining and decolonizing geography. Geografiska Ann: Ser B Hum Geogr 88(3):267–276
Short D (2016) Redefining genocide: settler colonialism, social death and ecocide. Zed Books, London
Silva AB (2007) Mais Além da "Aldeia": Território e Redes Sociais entre os Guarani de Mato Grosso do Sul. PhD thesis, Museu Nacional/UFRJ, Rio de Janeiro
Smithers Graeme C, Mandawe E (2017) Indigenous geographies: research as reconciliation. Int Indigenous Policy J 8(2). http://ir.lib.uwo.ca/iipj/vol8/iss2/2, https://doi.org/10.18584/iipj.2017.8.2.2
Thompson EP (1993) Customs in common. New Press, New York
van Krieken R (2004) Rethinking cultural genocide: Aboriginal child removal and settler-colonial state formation. Ocean 75(2): 125–151
Vietta K (2013) Histórias Territoriais: A Privatização das Terras Kaiowa como Estratégia para a Guarnição da Fronteira Brasileira e Outras Histórias. Espaço Ameríndio 7(2):26–63

Villar D, Combès I (2013) La Tierra Sin Mal. Leyenda de la Creación y Destrucción de un Mito. Tellus 24:201–225
Viveiros de Castro EB (1986) Araweté: Os Deuses Canibais. Jorge Zahar, Rio de Janeiro
Viveiros de Castro EB (1996) Images of nature and society in Amazonian ethnology. Annu Rev Anthropol 25:179–200
Woolford A, Thomas J (2017) Genocide of Canadian first nations. In: Totten S, Hitchcock RK (eds) Genocide of Indigenous peoples. Routledge, London and New York, pp 61–85
Yiftachel O, Fenster T (1997) Frontiers, planning and Indigenous peoples (Introduction to a Special Issue). Prog Planning 47:251–258
Zelic M (2018) Um Genocídio em Curso no Mato Grosso do Sul. Cartografia dos Ataques Contra Indígenas (CACI). http://caci.cimi.org.br

Chapter 8
Development and Conservation Frontiers in the Pantanal Wetland

Abstract The challenges related to the conservation of the Pantanal are analysed through a study of the Cuiabá River Basin, one of the most impacted areas in the Upper Paraguay River Basin and located in the Brazilian State of Mato Grosso. The research findings show a clear disconnection between official monitoring data and wider perceptions of processes of eco-hydrological change. In addition, it is found that putting environmental regulatory tools into practice is problematic because of the pregiven foundations of conservation strategies promoted by public agencies. The most significant result of the analysis is the uncertainty in relation to the responsibility for environmental problems, which has created a perverse chain of 'otherness–noneness–nothingness'. This serves to conceal the underlying causes of environmental degradation in the Cuiabá basin and limit possibilities for resolution. The main conclusion is that, beyond narrow development and conservation debates, it is necessary to account for a range of highly politicised issues at the intersection between interpersonal relations and socio-economic pressures.

Keywords Cuiaba · Environmental conservation · Consciousness · Mobilisation

8.1 Introduction

The Pantanal is one of the largest and most important tropical wetlands on the planet, as increasingly recognised in academic publications, international documents and government policies. It is located in the centre of the Upper Paraguay River Basin (UPRB) and has a total area of 147,574 km^2 shared between Bolivia, Brazil and Paraguay (ANA 2005). The Pantanal constitutes both a 'biodiversity hotspot' and a priority region for environmental conservation due to mounting threats caused by the expansion of agribusiness and engineering infrastructure (Olson and Dinerstein 2002). Around 80% of the Pantanal wetland is within Brazilian borders, which has particularly suffered from the consequences of urban and agro-industrial expansion in areas higher than 200 m in altitude (the plateaus that surround the Pantanal), as well as from the intensification of mechanised agriculture and the use of chemical in the floodplain (da Silva and Girard 2004). Changes in the plateaus and in the floodplain

A. A. R. Ioris, *Frontier Making in the Amazon*, Key Challenges in Geography,
https://doi.org/10.1007/978-3-030-38524-8_8

tend to reduce the variability in natural flow regimes and affect seasonal inundations (normally described as 'flood pulse'), which are critical for the maintenance of wildlife in the Pantanal (Hamilton 2002).

Accelerating rates of environmental degradation create a high level of uncertainty about the future of the Pantanal wetland system. The tension between available resources and mounting development pressures has led to recurrent calls for ecological conservation. At the same time, the management of ecosystems and natural resources is a matter of significant disagreement among social groups. The result is a situation with contradictory demands and reactive responses that are, at best, only marginally successful. Most of the conservation debate is still focused on a high-level description of problems that, in the end, fails to deal with the underlying socioecological complexity. The available interpretations concentrate on the disturbance of ecosystems, regulatory failures and lack of investments (see below), but it is rare to find critical analyses that connect conservation measures with personal subjectivities and socio-spatial inequalities. What is worse: for the majority of politicians and policymakers, conservation is typically seen as secondary to economic development in the whole region and agribusiness production in the plateaus.

The aim of this chapter is to enrich the debate about conservation alternatives for the Pantanal by incorporating comments, reflections and expectations of different groups of stakeholders along the lines of the conceptual formulation on frontier making introduced in the previous chapters. What follows is based on the results of fieldwork carried out in the Cuiabá River Basin, in the northern part of the Brazilian Pantanal—within the state of Mato Grosso—which is one of the areas with considerable levels of socio-economic activity and serious environmental risks. Although the Pantanal wetland is not formally inserted in the Amazon, it shares many ecosystems and biotic communities with that biome (as well as with the Atlantic Forest and Chaco forests and savannahs) and, more importantly, has been affected by the same driving forces of frontier making under the ideology and the agenda of regional development and national economic integration. The analysis builds upon two decades of the author's experience in project management and policymaking (working for the Brazilian Ministry of the Environment) and the coordination of an international scientific network on the Pantanal (see Ioris 2004, 2012). Although the empirical results provide only a snapshot of views and perceptions, these represent practical insights into the existing and future capacity to answer to emerging environmental conflicts. Before dealing with the specific circumstances of the Cuiabá River Basin, the next section will discuss the shortcomings of the mainstream conservation debate in the UPRB. That will be followed by the results and main findings of the case study in the catchment and, finally, by some general conclusions.

8.2 Conservation of a Global Wetland of Local Importance

The Pantanal is a wetland internationally famous for its lavish biodiversity, unique ecological features and cattle ranching traditions (Junk et al. 2011). The region has important economic activities, a strategic geopolitical location and abundant natural resources (land, water, minerals, biodiversity, etc.). As a result, the international community, including academics, diplomats and NGO activists, has emphasised the relevance of the Pantanal as a local wetland of global importance (often making reference to the Pantanal Matogrossense National Park, designated in 1981 and later declared a Ramsar site in 1993, and to the fact that the whole region was inscribed on the World Heritage List and designated a UNESCO Biosphere Reserve in the year 2000). However, the emphasis on the 'global importance' of the Pantanal has sometimes the perverse effect of concealing other aspects of its socioecology and the agency of local groups. There is a systematic failure to recognise the Pantanal as a global wetland of local importance, in other words, its geographical uniqueness should play a more important role in the understanding of problems and in a more inclusive discussion about conservation priorities. The long-term prospects of the Pantanal, and of the entire UPRB for that matter, actually depend on this intricate dialectic between global and local, as much as between conservation and development. It is crucial to consider the many links within socioecological systems and with other external processes in order to inform governance institutions (Anderies et al. 2004).

One problem with the current debate is the prevalence of biological measurements and quantitative, analytical experiments, at the expense of interdisciplinary approaches that could more adequately address social and spatial tensions mediated, and configured, through the interaction between society and the rest of nature. For instance, the Pantanal Agriculture Research Centre (CPAP), based in Corumbá (the main urban centre in the Pantanal floodplain), has an impressive team of ecology researchers but only a handful of social sciences experts. Likewise, the most comprehensive assessment of the Brazilian Pantanal to date, the Conservation Plan of the Upper Paraguay River Basin (PCBAP), suffers serious flaws that reflect the prevailing conservation rationale. PCBAP (1997) is a true 'encyclopaedia' of the Pantanal, a compilation of the best scientific information available at the time. In a series of volumes, it enlists environmental problems such as soil erosion (throughout the river basin and with particular severity in some specific locations), agriculture mechanisation (leading to soil degradation and sedimentation), pollution and devastation caused by wildcat miners (*garimpeiros*), deforestation and inadequate use of soil (in farmland and in riparian areas) and the use of agriculture pesticides. Yet, the PCBAP is largely an over descriptive and fragmented document that soon after being published began to accumulate dust on the shelves of academics and policymakers. A particular limitation of PCBAP is the persistence of the unresolved dilemma between environmental conservation and economic growth in a context of growing pressures for production intensification and income generation.

Other publications have acknowledged the exuberant eco-hydrology of the Pantanal, but did little more than registering the perceived main threats (i.e. water pollution, loss of biodiversity, mining, erosion and sedimentation, river regulation projects and modification of natural cycles) and superficially suggesting alternatives such as ecotourism, traditional cattle raising and the continuation of on-going projects (e.g. Arts et al. 2018; Cordeiro 1999; Hamilton 1999; Pott and Pott 2004; Rabelo et al. 2017; Swarts 2000). Similarly, Alho et al. (1988) express serious concerns over the removal of the native floodplain vegetation to make space for the introduction of artificial pastures while Alho (2011) suggests that the main problems are deforestation, water pollution, uncontrolled infrastructure expansion, unregulated tourism and the introduction of exotic species, but very little is said about the socio-economic disputes that underpin those trends. A series of studies commissioned by the Brazilian government, with international support, also aimed to ascertain the causal connections between problems and eco-hydrological impacts (ANA 2005), but the final product again fell short of considering the underlying causes and synergies between socioecological problems. Junk and da Cunha (2005) claim that low human population density and extensive cattle ranching in place since European colonisation had a little environmental impact. For the last authors, the Pantanal is essentially at a 'crossroads' due to mounting developmental pressures on ecosystem functions, which include the construction of roads and dams, as well as water pollution and overgrazing.

Despite the importance of those more recent analyses, major shortcomings persist in the understanding of intersectoral and multi-scalar connections that affect the conservation of the Pantanal. It still remains a 'great divide' between social and biophysical theories while there is a need for new methodological and interpretative approaches able to capture the full socioecological basis of environmental problems (Goldman and Schurman 2000). Socioecological systems like the Pantanal need to be seen at the interface between global and local phenomena, where such multi-scalar interconnections are a central feature of their vulnerability, resilience and adaptability (Young et al. 2006). The majority of the existing publications have paid limited attention to emerging disputes between urban and rural, old and new, landlords and employees, institutions and citizens regarding the access to natural resources and the asymmetric impacts of environmental degradation. The consequence is that policy-making has demonstrated an unwillingness to challenge conventional environmental regulation and fail to recognise the interlinkages between regional development and national and international economic failures. Coordination between government initiatives has been poor, with inadequate or partial geographical coverage, and has emphasised activities of low added value and high economic impact, such as farming, extractivism and mining (Tocantins et al. 2006). Also the new forest code approved by the National Congress in 2012 is expected to reduce the legal protection of riparian areas due to changes in the consideration of the water level of reference (Piedade et al. 2012). Similarly, the draft of a dedicated Pantanal Law (bill number as PL 750/2011), introduced in the National Congress in 2011 by the then Senator Maggi (later Secretary of State for Agriculture), one of the main leaders of the agribusiness sector, shows a distinct Malthusian bias when penalises, first of all, traditional communities living in the Pantanal.

There has been a recurring hesitation to tackle the internal contradictions of environmental policies and the politicised basis of environmental management responses. Most recent commentators ignore the demands of an increasing number of players, apart from the traditional cattle ranchers, miners and plantation farmers, involved in the disputes about the priorities of regional development, such as environmentalists, landless groups, family farmers, indigenous peoples, navigation companies, tourism operators and energy suppliers (i.e. natural gas from Bolivia to Brazil and hydropower dams being built at the transition between the plateaus and the floodplain). That means neglecting that socioecological systems are inherently reflexive and that, because of this reflexivity, initiatives aimed at avoiding or mitigating environmental dangers can be found (Young et al. 2006). On the one hand, the focus on ecosystem-based approaches has become more evident as the integration of scientific knowledge is required for the development and application of public policies. On the other hand, there has been no appetite among politicians and policymakers in the region to question the fast rate of agribusiness expansion (i.e. capital intensive and mechanised agricultural for commercial purposes) in the plateaus surrounding the floodplain.

The absence of more critical interpretations of the multiple interconnections between state, society and the rest of nature has led to a largely technocratic tone of conservation plans, which have reproduced a superficial understanding of the causes and consequences of environmental impacts. In the end, the strategies developed for the conservation of the Pantanal have operated within the same narrow episteme of regional development policies introduced in previous decades. There has been limited methodological and conceptual innovation and only a modest search for alternatives that are genuinely able to reconcile environmental conservation and the needs of the majority of the population. PCBAP (1997), for instance, includes mainly the outcomes of environmental disruption rather than dealing with the underlying drivers of socioecological changes and with the perpetuation of inequalities. Likewise, environmental regulatory reforms have been largely restricted to changes in the structure of government agencies and environmental legislation but allocated an insufficient effort in terms of regulation enforcement and the democratisation of decision-making. Safford (2010) shows that, while the new water legislation delegated to catchment committees the approval of plans and the reconciliation of spatial differences, in practice there has been only partial regulatory enforcement and the maintenance of long-established administrative procedures.

Although the sustainability of the Pantanal was incorporated into legislation, policies and official discourses, the crucial decisions about the economy, infrastructure and public services continue to follow the wider balance of power and, in particular, the hegemonic interests of the agribusiness sector. Even if announced by public agencies as something neutral and universally advantageous, contemporary environmental policies do not leave room for the long-term, politicised interactions between different social groups mediated by the access to (the rest of) nature (Ioris 2010). Instead of promoting a genuine change in public policies, prevailing approaches have largely preserved the interests of landowners, industrialists, construction companies and real estate investors, at the expense of the majority of the population and the recovery of ecological systems. For instance, there is a repeated attempt to associate public–private partnerships with novel responses to environmental degradation. TNC-WWF (2011: 13) specifically endorse the role of financial institutions 'to incorporate environmentally sustainable requirements' when extending credit to agriculture and cattle ranching in the Pantanal. Those new tendencies raise serious questions about the legitimacy of the more recent policies and programmes formulated for the region, especially because they reproduce the same pattern of unequal, short-term results of traditional development (see the case study below). That is a clear demonstration of what Büscher et al. (2012) call 'neoliberal conservation functions' that serve to further entrain nature to capitalism and to create broader economic possibilities for capitalist expansion.

The problem is not simply the lack of legal structures and institutions aimed at limiting the environmental impacts of rapid development and land-use change (i.e. there are actually more than 120 laws on environmental issues connected to the Pantanal and its surrounding areas alone, as discussed by Charnoz 2010), but environmental degradation systematically replicates the politics of regional development. If a violation is detected, it is often not enforced, the penalty is not large enough to act as a deterrent against non-compliance or there are cases of corruption involving law enforcers and environmental aggressors (Meio Ambiente Agora 2011). The old plan to extend the Paraguay–Parana Waterway (*hidrovia*) also illustrates the difficulty to go beyond the narrow basis of the conservation-development debate (Gottgens 2000; Gottgens et al. 2001; Schulz et al. 2017). The project was shelved in 1998 but then brought back to public attention after its inclusion in the Programme for the Acceleration of Growth (PAC), the national agenda of investment by the Brazilian government launched in 2007. However, the debate between agribusiness, river engineers and environmentalists has largely ignored the needs of the riparian communities living along the Paraguay River, which are increasingly sceptical of initiatives that in the end reinforce patterns of social inequality and environmental degradation (Borges et al. 2000). An emerging threat is the aggressive expansion of hydropower generation, which has become one of the main environmental pressures on the Pantanal, particularly due to the growing demand for energy in the industrial areas of Brazil. The increase of hydropower (around 140 new hydropower schemes are under construction or being planned in the Brazilian side of the UPRB), together with the large areas with sugarcane (to produce ethanol), accentuates the tensions between contrasting scales of the environmental agendas, in this case the reduction of greenhouse gas emissions to cope with climate change and the local impacts caused by dams and intense farming (Ioris et al. 2014).

The above publications and projects reveal the difficulty to reconcile the global importance of the Pantanal and the promotion of concrete action measures. The biased interventions of the state apparatus, including the work of environmental regulatory agencies, have primarily supported the expansion of urban growth and agribusiness activities. The impacts of socio-economic growth in and around the Pantanal floodplain in recent decades—mainly associated with the export of agriculture commodities—can only be addressed with responses that are largely beyond the existing scientific approaches and the commitments of existing policymaking frameworks. Scientific and regulatory uncertainties have not been helped by the lack of integration between researchers and academic communities, but there are more basic questions still to be addressed. There is a pressing need to unpack the underlying barriers to effective conservation of the Pantanal, which calls for a broader consideration of the connections between governmental, sectoral and interpersonal synergies. In that context, the alteration of river basins, like the Cuiabá River, represents a telling example of socioecological changes, environmental degradation and lasting inequalities in the Pantanal region.

8.3 The Cuiabá River Basin as a Microcosm of Pantanal's Conservation Dilemmas

This section presents the results of the case study in the Cuiabá River Basin, which is one of the largest, most populated and extensively impacted catchments in the UPRB. The local experience is highly illustrative of the contemporary conservation challenges and serves as an entry point into the wider—and necessarily more complex—questions of socio-economic development in the UPRB. The river basin has an area of 28,732 km^2 that can be schematically divided into three main sections: the plateaus with intensive plantation farming, the medium section around the city of Cuiabá (the capital of Mato Grosso) and the Pantanal floodplain (Fig. 8.1). The annual average temperature is 26.8 °C and average precipitation is between 1,700 (in the headwaters) and 1,300 (in the floodplain); average river flow oscillates between 300 and 350 m^3/s (data from the Federal University of Mato Grosso—UFMT). It is beyond the scope of this chapter to describe in detail the Cuiabá River Basin, but specific information about the river basin and the Pantanal at large can be found, among others, in Ecoplan (2003), Figueiredo et al. (2018), Zeilhofer et al. (2010), Junk et al. (2011) and Ioris (2012).

The river basin is shared by 14 local authorities and has a population of around one million with an urbanisation rate of 93% (Ecoplan 2003). Investments in water services have lagged behind the rate of urban growth and only 30% of the urban sewage is collected and treated (Sanecap 2011). At the time of the research there was an expectation of additional funds to prepare the city for the 2014 World Football Cup, but evidences of widespread corruption seriously affected the construction timetable. As a result, the river system, which has historically been one of the main recreation options for the locals, was then banned by the municipalities downstream of the city of Cuiabá due to the increasing contamination by faecal coliforms and other forms of pollution (Figueiredo 2009). In addition, a large hydropower scheme—the Manson dam—entered into operation in 1999 and impounds water of the Rivers Manso and Casca, tributaries of the main Cuiabá River. This complex combination between long-established uses of the river and new management approaches provided the context and the justification for the analysis of the underlying barriers to environmental conservation.

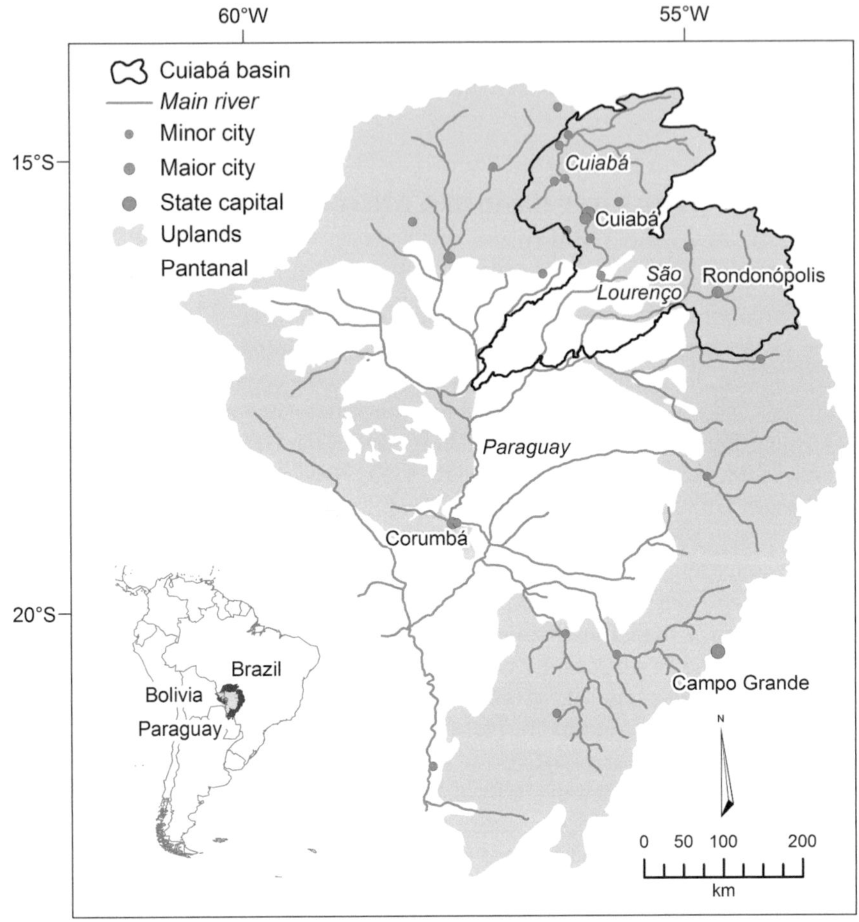

Fig. 8.1 Map of the UPRB, the Pantanal and the Cuiabá River Basin

8.3.1 Methodological Approach and Research Findings

The investigation consisted of an exploratory study designed to capture and compare underlying values, perceptions and expectations about the future. Different than most existing assessments mentioned above, the current analysis employed a bottom-up approach. The methodology combined interviews, participant observation during site visits and the attendance of public events for a period of eight months in 2011. The overall purpose was to determine how the local knowledge is constructed and comprehend the different views about pressures affecting the Pantanal and the Cuiabá River Basin in particular (Fig. 8.2). An initial list of stakeholders was selected in different places along the Cuiabá River, based on the suggestions made by local academics at UFMT. Further contacts were identified following a snowball approach

Fig. 8.2 Cuiabá River near the city of Cuiabá

(Bernard 2002) where respondents gradually provided the names and contact information for other social actors (Babbie 2001) living in the same locality. Interviews (35 in total) were conducted in two fieldwork campaigns: 18 in March and 17 interviews between May and July. The basic criteria to plan the interviews were to include locations upstream and downstream to the city of Cuiabá, as well as the metropolitan area, in order to incorporate as many stakeholder sectors as possible, particularly those not usually included in the formal decision-making process. Stakeholders from the following sectors were contacted as follows: (a) professional fishermen, (b) cattle farmers and representatives of the farming sector, (c) water utility officers, (d) park rangers, environmental officers and environmental guides, (e) recreational fishermen, (f) manufacturing industries, (g) owners of restaurants and tourism agencies and (h) public authorities and legislators. Interviews were semi-structured and included questions defined in the earlier phases and informed by the available academic and non-academic literature.

The study followed an interactive strategy that tried to compose a synthesis of multifarious processes associated with water use and environmental management in a specific geographical context. According to the inductive nature of the research, the explanation was neither objective nor neutral, but intrinsically connected to the personal experience of the researcher and, therefore, the interpretation of processes under consideration. Instead of predetermined hypothesis, the study focused on a list of questions that allowed the interviewees to provide in-depth responses and make

Table 8.1 Key themes and insights from the interviews

Key themes (i.e. NVivo nodes)	Proportion in the interview transcripts (%)
Critique of environmental problems	19.6
Demands of public services	10.9
Environmental values	6.5
Details of water uses	13.0
Disputes and conflicts	15.2
Collaboration and alliances	8.7
Public policies and politics	26.1

subjective assessments (Ioris 2014). A set of ten open-ended questions were used in the interviews in order to encourage the participants to shape their own narratives of the lived experience in the river basin, their understanding of problems and demands in terms of environmental restoration and conservation. To allow for depth, nuance and to acquire information on its natural form, the interviews (which lasted around 1 h each) were made face-to-face, taped, transcribed and analysed making use of the NVivo software.

For the use of NVivo, seven main nodes and 25 sub-nodes were determined based on the interview questions and on a preliminary assessment of the interview transcripts; once the nodes and sub-nodes were available, the full text of the transcripts was analysed and comparable answers and claims were grouped accordingly. Table 8.1 has a summary of the main interview themes (converted into main NVivo nodes) and the proportion of their appearance in interview transcripts. Note that it will be presented here as the most relevant, instructive findings of the research concerning environmental conservation and management. The empirical results—considering the association of nodes and sub-nodes—suggest that the three main obstacles of environmental conservation in the Cuiabá River Basin are the conflicting perception of problems, the rigid formulation of responses and the vague responsibility for environmental degradation, which will be discussed next.

8.3.2 *Conflicting Perceptions of Socioecological Trends*

One of the main objectives of the research project was to contrast the opinion of different social groups with scientific publications and the rationale of public policies. Academic and governmental documents vaguely acknowledge the trend of alterations, especially those associated with fast urbanisation in the middle stretches of the river and agribusiness and dam construction in the upstream section (Mato Grosso 2009). Such pressures are described as disrupting the eco-hydrology of the river and aggravated by overfishing and deforestation. Water quality surveillance is carried out by the state environmental protection agency (SEMA), basically making use of methodologies developed in northern countries and adjusted to the condition of tropical catchments. Nonetheless, the official monitoring network has still limited statistical evidence of environmental degradation and water pollution. Out of the only 15 sampling points used by SEMA along the Cuiabá River, water quality is classified as good in the majority of points during most of the year; water quality deteriorates to a medium condition in certain months due to lower oxygen, higher concentration of coliforms and some other pollutants from urban and rural diffuse sources (see more details in SEMA 2010). There is better evidence of water quality degradation only in the area in and around the city of Cuiabá, where the river is significantly impacted by the lack of sanitation infrastructure, the connection with the storm water system and the expansion of impervious surfaces (Zeilhofer et al. 2010).

It is well-known by the local academics and policymakers that the coverage and frequency of water quality monitoring in the Cuiabá River Basin is far from the level recommended for large tropical catchments. What is less often discussed is that the process of environmental change has not been equally perceived, and acted upon, by different communities and social groups. The empirical results of the current study revealed a surprising, and growing disparity in the understandings and in the reactions to the poor environmental condition of the catchment. More importantly, there was a noticeable mismatch between the reaction of most of the interview respondents and the water quality problems recorded by the monitoring agency (obviously it is the latter, instead of the former, that inform the management procedures adopted by environmental regulators). Especially the population living close to the river or with some regular contact with the water system sustain a more negative impression about the biochemical deterioration than the picture described in official reports. But all respondents who took part in the research expressed strong views about the trend of impacts and complained about what they see as an accelerating environmental degradation and the inadequacy of most public policies, for example, 'The main problem is the apparent lack of public policies dedicated to the conservation and preservation of the river. [We need] more focused and effective policies' (interview, resident of Santo Antônio do Leverger). See Fig. 8.3.

When asked whether water could be consumed directly from the river, almost all respondents categorically refused to contemplate the idea of drinking untreated water, which is a vivid indication of the underlying perception of the status of the river system, as is illustrated by the following quotation:

> I would not drink. Because we notice, when we collect some water on a jar, that is no longer just water, it is something cloudy, murky. There is so much left in the water. Water is no longer as it used to be. It is a different type of water. (…) Nowadays we use it to wash something, but never to drink. Before we use we boil it. But we no longer drink water from the Cuiabá River. Especially in the dry season [middle of the year], when the river is low, we see very clearly the pollution as a white grease. (interview, fisherman, São Gonçalo Beira Rio community)

This gap between scientific and non-scientific assessments of water quality is not a trivial difference between expert and lay interpretations. On the contrary, it exposes a considerable distance between the official treatment of risks and uncertainties and the wider social perception of the processes of change affecting the river

Fig. 8.3 Cuiabá River at Santo Antônio do Leverger

system. Although the public is unable to quantify the alteration in numeric terms, in their opinion and based on their lived experiences the river is increasingly becoming more degraded. Instead of dissimilar discourses, these represent entirely different 'grammars' of environmental conservation used by social groups and organisations. It is a diversity of environmental 'grammars' in the sense that each one denotes specific symbols, values and attitudes towards the environment. Distinct environmental 'grammars' are demonstrated by the contrast between, for example, the discourse of cattle ranchers, industrialist and hydropower operators, on the one hand, and fisherman, riparian communities and low-income residents, on the other. The plurality of environmental 'grammars' found in the Cuiabá River Basin actually serves as a representative example of the confusion about the conservation and management of the Pantanal at large. The contrasting opinions about the status of the catchment are held according to the ability to influence policymaking and regional development. 'Grammars' are not only dissimilar but reflect a whole hierarchy of power that underpins the various claims about the river system. The interrelationship between knowledge and power is evident in the 'grammar' used by the stronger water users (public and private), who feel powerful enough to defy both public criticism and scientific advice (i.e. stronger economic groups are normally associated with sectors that significantly alter the river system, such as agriculture, mining and industrial production, and who systematically strive to contain the economic impact of a more stringent environmental regulation).

The construction and operation of the Manso Hydropower Scheme provide a clear demonstration of conflicting 'environmental grammars' underpinning perceptions and reactions to environmental change. The initial environmental statement identified a range of likely environmental impacts and included 21 monitoring and mitigation programmes, such as hydrological, limnological and water quality surveys (details are available in FURNAS 2012). However, since the construction of the dam, mounting environmental problems have been denounced by riparian residents, but systematically disputed by the hydropower operator FURNAS. The very enforcement of the environmental regulation by the environment protection agency (SEMA) has been problematic and resisted by the managers of the hydroelectric scheme. Lenient regulation has not been helped by the fact that FURNAS is an agency of the federal government (i.e. formally beyond the remit of SEMA) and the enhancement of electricity supply is one of the key infrastructure priorities of national development policies. At the same time that this controversy about the primary regulatory responsibility remains unsettled; almost all individuals interviewed during the research expressed substantial reservations about the Manso dam and what they perceive as significant changes in the season flow regime and the associated decline of fish populations. Many respondents affirmed that the river flow is now disturbingly different than used to be in the recent past:

> What we can clearly see is that water has reduced due to the existence of Manso today. The river used to 'come' up to the top of the riverbank in periods of heavy rain. But now is only reaches a certain point, never overflows the riverbank. (…) And the quality is also different; here in our community we don't even enter into the water. (interview, artisan, São Gonçalo Beira Rio community)

In the end, there exists uncomfortable anxiety among many sectors of the local society about indeterminate, but palpable, water quality and water quantity alterations. The public may be unable to measure the level of environmental risk, but they are convinced that the river has been affected by the urban expansion and economic activity in the catchment. Those concerns are intermingled with a sense of nostalgia about the past condition of the Cuiabá River system in previous decades. A perverse consequence of the overall ambiguity about the changes happening to the catchment is the alienation, and even exclusion, of traditional water users from the actual management of the river system. The population shows little tolerance with a situation that is seen by many, even in an imprecise way, as unfair restrictions to the access and use of the river. In the interviews, some respondents related the growing rate of environmental impacts in the catchment with what they perceive as a condition of serious inequality in terms of access to the river and sustained disregard of their 'lay' opinions. For instance, professional fishermen complained about what they see as the privatisation of the margins of the river by restaurants and hotels, which introduce fishnets and other physical structures to contain the fish only for their own interests. That was vehemently criticised as an unjust practice that further reduces the income of activity already struggling to survive in the Cuiabá River Basin. While the water quality seems to be deteriorating fast and affecting many social groups, wealthier people are allowed to build large mansions (normally without proper planning permission) along the riverbanks:

> There should be a more stringent control of activities in the catchment. If you take a boat and travel down the river you will see such magnificent houses, including some that belong to wealthy judges [a profession normally associated with a privileged social status in Brazil], who even build little weirs in from of their houses. These people, instead of giving a good example, they think they can do whatever they want. (interview, community leader, de Nossa Senhora da Guia association)

The series of cleavages in the discussion about the actual condition of the river and the future of the catchment serve to reinforce an overall situation of environmental degradation and socio-spatial inequalities. On the one hand, there are calls for action by governmental and non-governmental organisations for dealing with an indeterminate, but clearly uncomfortable, situation. On the other hand, there is a persistent difficulty among the general public to connect causes and effects, which is further encouraged by hesitant scientific assessments and the sociopolitical commitments of the state. That ends up producing a sort of 'conservation paralysis', that is, a major difficulty to put in practice legal requirements and public policy goals. Habermas (1987) makes reference to the inherent dialectic of modernisation as the burdens placed on the internal structures of the lifeworld by growing system complexity. This well-known tension of European modernity continues to operate in areas recently incorporated into the globalisation of markets, as in the case of agribusiness intensification and mass commodity consumption such as the Cuiabá River Basin. It is important to realise that such a dialectic of modernisation functions not only at the macro level of regional development and formal arenas of political disputes but also through the intricacies and non-linearities of everyday life and interpersonal dealings. The convergence between wider system complexity and the everyday lifeworld has significant repercussions for the comprehension and management of environmental processes, which leads us to the next main shortcoming of environmental conservation.

8.3.3 Pregiven Responses to (Uncertain) Socioecological Problems

There are main consequences of the persistent disregard for what is considered an emotional, misinformed reaction of the local population to the environmental problems of Cuiabá catchment (as discussed in the preceding section). First, the conflict between the official discourse and a myriad of public opinion on different issues has resulted in the lack of environmental conservation leadership, particularly by the state water agency (SEMA) and the Mato Grosso water council (CEHIDRO). Second, there is a related difficulty to put into practice the growing number of legal norms and environmental regulatory tools. Our empirical results suggest that the persistent difficulty to advance environmental regulation derives primarily from the external, pregiven foundations of most contemporary conservation strategies. Formal environmental rules have expanded significantly in the last two decades, but the core elements of the new legislation—such as environmental impact assessment, user permits and charges, and the payment for ecosystem services (Schulz et al. 2015)—have more to do with the circumstances in the south-eastern parts of the country than with the particular geography of the Pantanal region (or the Amazon and the Cerrado, for that matter). The Mato Grosso water law, for example, follows the tenets of the national water law (both were passed in the same year 1997) and the two incorporated the international doctrine of integrated catchment management, techno-economic instruments of regulation and representation in catchment committees (Ioris 2009).

The new water legislation is centred on the need to pay for the extraction of water from the environment (in order to increase use efficiency), as well as on the formation of river basin committees (supposed to be in charge of resolving water conflicts). Yet, those instruments seem largely inappropriate for a situation where most forms of water use are not consumptive (i.e. the number of water users that abstract water from the river, such industries and irrigation farms, is very low in the Cuiabá catchment) and the water management controversies are beyond the remit of state authorities (as in the aforementioned case of the Manso dam). The result is that both water charges and the committee are still not operational and have restricted prospects to derive actual socioecological benefits to the catchment. One important reason for the still inexistence of the river basin committee is the enduring dispute between the government of Mato Grosso and the National Water Authority (ANA) about the legal responsibility for the catchment (i.e. formally, it is a federal river basin but in effect its almost entirely contained in the state of Mato Grosso).

It was specifically mentioned in some interviews that adjustments in the organisation of public agencies and in the institutional arrangements have not corresponded to the mounting environmental management problems in the Cuiabá River Basin. On the contrary, the prevailing public perception is that the responses of government agencies have been limited, tardy and given rise to populism. Most initiatives fail to encompass the geographical particularities of the catchment in favour of targets and objectives that are only partially connected to the wider universe of public opinions and the intricacies of local socio-economic activities. In that sense, the lack of scientific data and the difficulty to comprehend the complexity of the river basin (repeatedly mentioned in publications like Alho 2011; Galdino et al. 2006; Lourival et al. 2008; Swarts 2000) becomes a convenient excuse for policymakers and scientists to avoid difficult questions about the soundness of natural resources management and environmental conservation frameworks.

In a context of generic, pregiven answers to idiosyncratic socioecological problems, other forms of community initiatives and grassroots organisation tend to be regularly ignored. In the same way that unofficial information about the river is disregarded by the formal environmental regulatory machine, bottom-up reactions to environmental problems are unapproved and unsupported by decision makers, as reflected in the following statement:

> Look, there are lots of those available [public policies], but the problem is that they don't work. Lots of policies, but they are beyond what we need. The solutions don't move [things] forward, while the problems keep mounting. (…) I believe that it is a clear lack of political

> willingness, because one has to incorporate different variables, work with several aspects simultaneously, but the government has no interest in doing that. (interview, NGO member)

The difficulty to engage with the public reveals a great deal about the weaknesses of environmental regulation in a situation of fast-expanding regional development. The conversion of the original landscape of the Cuiabá River Basin into plantation farms and the intensification in the use of natural resources constitute the prevailing vectors of economic growth and, as a result, there is limited political will to adopt any environmental regulation measures that seriously clashes with those hegemonic interests (as demonstrated by the introduction of the bill PL 750/2011 mentioned above). The same government that is responsible for environmental conservation and scientific research is simultaneously the main promoter, and beneficiary, of such a conservative model of economic development that takes place in the upland areas surrounding the Pantanal, where agribusiness and intensive agro-industry have appropriated large extensions of land and resources. Instead of internal coherence and good coordination between scales of government, the state apparatus is permeable and especially susceptible to the influences of the most powerful social groups (Ioris 2010). And it is also evident that the environmental branch of the Mato Grosso government (i.e. SEMA) is one of the politically weaker and less important divisions of the state apparatus.

Interestingly, although the primary commitment of the Mato Grosso government remains the expansion of crop production, the social imaginary of the Pantanal—populated by colourful images of birds, fish and water—has been consistently appropriated by politicians as supposed evidence of their environmental credentials. The fact that it is still possible to find in the Pantanal ecosystems still relatively well preserved (what is the result mainly of the difficult access to most of the floodplain rather than the consequence of conservation initiatives) effectively operates as an element of the legitimisation of hegemonic development policies. As much as the symbolism of the Pantanal operates as the 'moral reserve' of public authorities in charge of environmental conservation, it is also constantly emphasised as one of the priority regions for the expansion of tourism and ecotourism in Mato Grosso (Fig. 8.4). However, in our interviews the local population and their leaders resented the uncontrolled growth of tourism in the Pantanal floodplain and along the rivers of the Cuiabá River Basin that are tolerated or stimulated by the state government. For instance, the mayor of the town of Barão de Melgaço expressed his frustration with the construction of a paved road to his municipality, which provided easy access to sport fishermen (blamed for additional environmental degradation and for bringing virtually no income to the local markets, as they arrive with their own equipment and provisions) without the necessary control by environmental authorities. The ordinary blaming of 'other' people and other social groups have more than trivial repercussions but constitutes another serious barrier to environmental conservation, as discussed below.

Fig. 8.4 Leaflet by the Mato Grosso State Government associating nature with tourism

8.3.4 The Responsibility of the 'Vague Other'

Probably one of the most significant results of the research was the uncertain reference, in various interviews, to the responsibilities for environmental problems. While there is a generalised perception that something is badly wrong with the management of the river basin, there is a great difficulty to locate blame for the growing rate of environmental impacts. Most interviewees seemed convinced that the environmental problems are growing at a fast rate, the liability is placed at an undefined 'other', a fluid entity that is essentially a hazy combination of governmental agencies, public authorities and certain members of the public. The responsibility is largely obscure and indeterminate; it is typically related to someone else, the 'vague other'. This imprecise 'other' is someone who, because of his/her/its faults hijacks the river from the rest of society, but can't be properly identified, as demonstrated in the following interview extracts:

> I believe that the pollution in our Cuiabá [River] is the people themselves. I mean… how I could say, people in general, you know… Those who throw rubbish in the water, and many other things, also in the tributaries [of the main river], like in the Coxipó River. (…) In the rainy season, with all that rubbish, you can easily see the amount of fish, little fish, dying because of the water, you know. (interview, fisherman 1, São Gonçalo Beira Rio community)

> There is a clear lack of conscience; people should stop throwing waste in the river. Sometimes we go and collect the rubbish along the riverbank and the quantity is not small… (interview, community leader, de Nossa Senhora da Guia association)

> All the fishermen plan it, we go downriver and clear the riverbanks, but it is not enough, because we do it and other fail to protect [the river]. (interview, fisherman 2, São Gonçalo Beira Rio community)

> Look, up to ten, I would give a mark of two, because there is no way we can use the water from the river anymore. Only to water some plants, that is the maximum we do nowadays. (interview, restaurant owner).

> In my opinion, there is a lack of no scruple among the public, these are things that should never happen. Forty years ago nobody had any preoccupation with the environment, everything was fine. Environmental awareness is something too recent in Mato Grosso, very new indeed. What is missing is the public to take action and fight for some improvement. (interview, tourism agent)

That may sound like an irrelevant question of misinformation, but there are important social factors that may lead individuals to project the responsibility of solving environmental problems on vaguely defined 'others'. First of all, it is an integral part of the wider problem of the disregard and alienation of riparian communities by other urban and rural groups. It is also an indicator of the distant, problematic relation between residents and their ecosystems under lifestyle changes mediated by mass consumption, mass waste and the monetisation of everything. Another dimension of the same problem is the difficulty to relate the deterioration of the Cuiabá River with the failure of other governmental interventions, such as inadequate urban planning, deficient public transport and the submission to the strong interest of agro-industrial sectors. All those factors are then translated into a sense of powerlessness, as if the only thing left was to imprecisely blame someone else.

Together with the romanticism of the past condition of the catchment, the public struggles to make sense of the changing condition of the river and the actual responsibilities for environmental problems. The 'vague other' is also a manifestation of the weakness of the democratic institutions, so far as the general population is continuously relegated to a paternalistic relationship with the public authorities. Hence the nebulous blaming and the ambiguous expectation that the government agencies should do 'something' about the river system. Such disparity between rising concerns over environmental degradation and a range of undetermined responsibilities is neither easily acknowledged in the environmental conservation discourse (which is typically focused on the consequences of environmental disruption and fails to establish a consistent connection with the basis of socio-economic development), nor in most political ecology literature (which usually criticises the negative consequences of economic development without necessarily addressing personal and interpersonal processes). If the other is everybody, in practice it becomes nobody. That leaves a subtle sense of noneness, an uncomfortable feeling of helplessness or nowhere to go. It creates a distressing chain of 'otherness–noneness–nothingness' that ends up concealing and even reinforcing the overall process of environmental degradation (i.e. consciously or unconsciously, it supports the rationale that 'if nobody is doing anything, so it is not my obligation to do anything either'). This vicious circle of vague otherness and no responsibility is more serious than the commonly described NIMBY phenomenon ('not in my backyard', or the opposition by residents to proposals or initiatives close to them, although generally needed in the society).

The general uncertainly about environmental trends and regulatory responses reflects the wider context of political demobilisation and civil society disorganisation in the Cuiabá River Basin. The strong symbolism of water abundance has become a foundational element of the collective identity in the region, situated in the wider context of natural resource abundance more generally, what makes water abundance a contested concept witnessing discursive struggles around its political implications and meaning (Schulz and Ioris 2017). The local experience has shown that it is easier to contain public criticism and divert attention away from the mounting environmental problems if and when the liability is not clearly understood. Environmental injustice goes beyond the asymmetric distribution of opportunities and negative impacts, but it is also a manifestation of the hierarchical and discriminatory treatment of popular demands and the paternalistic interaction with the population. Furthermore, the combination of the previous two processes (i.e. contrasting perceptions of the catchment condition and the imposition of pregiven solutions) with the 'vague otherness' produce a widespread pessimism about the future of the river and its ecological status. As observed by some local residents:

> Unfortunately, everything seems to be worsening: and what is worse, I can't see a good future [for the river]. I can't notice any improvement. If you go and check on GoogleMaps, the deforestation in the catchment is growing and this is one of the critical problems. The tendency is only to go downhill. (interview, fisherman, Nossa Senhora da Guia community)

> It is a calamity to pay R$ 5.00 [US$ 2.50 at the time] for a large bottle of [mineral] water. (…) In remote areas we still have protected springs and there are wells with good water quality. But in thirty years, I believe that there will not be good water even to have a shower, for our personal hygiene. That is because of the unchecked development and lack of sanitation inspection. (interview, restaurant worker)

That sense of pessimism is constantly nourished by reports about the environmental degradation of the catchment that is disseminated by the mass media without a proper discussion of causes, consequences and responsibilities. An emblematic demonstration of the useless pessimism about the Cuiabá River was the comparison, in several interviews, with the Tietê River, which crosses the city of São Paulo and is famously one of the most polluted water bodies in the country:

> I don't believe that we could fish in the future in the Cuiabá River. I know the entire river, almost the whole Pantanal and I know how it is. You asked me about the changes in the last twenty years, but I want to answer that in the next thirty years we will have nothing left. I am convinced that we will only have a new Tietê here in Mato Grosso. (interview, fisherman, Z1 association)

8.4 Lessons Learned and the Prospects for the Pantanal

The previous pages briefly discussed the results of a study carried out in the Cuiabá River Basin, which provides an entry point into understanding the wider complexity of the challenges facing the Pantanal wetland. The main objective of the study was to investigate the connections between personal perception, public policies and future expectations about the environmental condition of the river basin. The initial review of the academic and grey literature on the Pantanal exposed the narrowly defined concerns of the existing public policies favouring diversity of social groups, conflicting political interests and supporting the legacy of past economic development. The conventional analyses of problems remain persistently superficial in terms of connecting the politico-economic basis of agribusiness and fast urbanisation with long-term environmental risks. In that politico-institutional context, the specific case study on the Cuiabá identified three main issues, which are relevant not only for the conservation of the Pantanal and tropical wetlands but also for environmental management in general: (1) the conflicting assessment of environmental problems and the contrasting perception of different social groups about who is to blame for them; (2) the inadequacy of existing policy responses imported from geographically different regions; and (3) the uncertain and vague allocation of responsibilities that happens through an unfortunate chain of 'otherness–noneness–nothingness'.

Overall, it was possible to clearly identify a widespread sense of uneasiness with both the condition of the Cuiabá River Basin and with what is seen as the inadequate responses of public agencies. The fact that the public normally holds opinions that contrast with the official discourse and the interests of strong stakeholder groups represents a major difficulty for the enforcement of regulation as well as for and the mobilisation of society. In addition, despite the existence of relatively comprehensive environmental legislation (such as that of the 1997 Water Law), the introduction and enforcement of environmental regulation have happened through a patchwork of measures that often are not consistent with each other and fail to lead to results in line with scientific claims, popular knowledge and the formal objectives defined for the Pantanal by government agencies. Finally, it was somehow surprising that the population both expressed discomfort with the performance of public agencies and persistently struggled to allocate and take personal responsibilities for the condition of the catchment. The responsibility is normally transferred to a 'vague other', someone distant and indeterminate.

Although not often recognised as such, those three synergic problems certainly represent challenging obstacles for the conservation of the river basin and the Pantanal and, what is more important, operate as a tacit justification of perverse economic development trends. The ultimate conclusion is that effective environmental protection of the Pantanal can only result if a critique of hegemonic socio-economic policies is productively connected with the complex perceptions and processes that determine personal behaviours and interpersonal exchanges. There is a crucial association between different scales of socioecological interaction—the national and regional with the local and personal—but it is still rare to notice its proper consideration in the conservation debate about the future of the Pantanal. However, such multi-scalar and highly politicised interactions cannot and should not be any longer ignored.

Acknowledgements This chapter is based on a previously published article by the author: Ioris, A. A. R. (2013). Rethinking Brazil's Pantanal Wetland: Beyond Narrow Development and Conservation

Debates. The Journal of Environment & Development, 22(3), 239–260. https://doi.org/10.1177/1070496513493276.

Reprinted with the kind permission of SAGE.

References

Alho CJR (2011) Biodiversity of the Pantanal: its magnitude, human occupation, environmental threats and challenges for conservation. Braz J Bio 71(1, suppl):229–232

Alho CJR, Lacher TE Jr, Gonçalves HC (1988) Environmental degradation in the Pantanal ecosystem of Brazil. Bioscience 38(3):164–171

ANA (National Water Authority) (2005) Strategic action program for the INTEGRATED Management of the Pantanal and the upper Paraguay River Basin: final report. ANA/GEF/PNUMA/OEA, Brasília

Anderies JM, Janssen MA, Ostrom E (2004) A framework to analyze the robustness of socialecological systems from an institutional perspective. Ecol Soc 9(1):18. http://www.ecologyandsociety.org/vol9/iss1/art18

Arts K, Rabelo MTO, Figueiredo DM, Maffey G, Ioris AAR Girard P (2018) Online and offline representations of biocultural diversity: a political ecology perspective on nature-based tourism and indigenous communities in the Brazilian Pantanal. Sustainability 10(10):3643. https://doi.org/10.3390/su10103643

Babbie E (2001) The practice of social research. Wadsworth Thomson, Belmont CA

Bernard H (2002) Research methods in anthropology: qualitative and quantitative approaches. Altamira Press, Walnut Creek

Borges JRP, Chaudhry FH, Ferreira Filho P (2000) Um Estudo sobre a Percepção de Ribeirinhos da Bacia do Alto Paraguai a Respeito da Questão de Justiça e Eqüidade em Aproveitamento Hídrico: O Caso da Hidrovia Paraguai-Paraná. Anais - III Simpósio sobre Recursos Natuais e Sócioeconômicos do Pantanal, Corumbá, 27–30 Nov 2000. CPAP-Embrapa

Büscher B, Sullivan S, Neves K, Igoe J, Brockington D (2012) Towards a synthesized critique of neoliberal biodiversity conservation. Capital Nat Social 23(2):4–30

Charnoz O (2010) Community participation in the Pantanal, Brazil: containment games and learning processes. AFD Working Paper No. 93. Agence Française de Développement, Paris

Cordeiro NV (1999) Environmental management issues in the Plata Basin. In: Biswas AK, Cordeiro NV, Braga BPF, Tortajada C (eds) Management of Latin American River Basins: Amazon, Plata, and São Francisco. UNUP, Tokyo, pp 148–173

da Silva CJ, Girard P (2004) New Challenges in the management of the Brazilian Pantanal and catchment area. Wetlands Ecol Manage 12:553–561

Ecoplan (2003) Diagnóstico da Bacia Hidrográfica do Rio Cuiabá: Relatório Final. Secretaria de Agricultura e Assuntos Fundiários do Estado de Mato Grosso, Cuiabá

Figueiredo DM (2009) Limnologia e Qualidade das Águas Superficiais das Sub-bacias Alto e Média. In: Figueiredo DM, Salomão FXT (eds) Bacia do Rio Cuiabá: Uma Abordagem Socioambiental. EdUFMT, Cuiabá, pp 114–125

Figueiredo DM, Calheiros DF, Ioris AAR (2018) A Bacia Hidrográfica: Integração do Conhecimento e Reflexão sobre o Desenvolvimento Sustentável. In: Figueiredo DM, Dores EFGC, Lima ZM (eds) Bacia do Rio Cuiabá: Uma Abordagem Socioambiental. EdUFMT, Cuiabá, pp 642–682

FURNAS (2012) Usina Hidrelétrica de Manso. http://www.furnas.com.br

Galdino S, Vieira LM, Pellegrin LA (2006) Environmental and socioeconomic impacts in the Taquari River Basin-Pantanal. Corumbá, CPAP-Embrapa

Goldman M, Schurman RA (2000) Closing the 'Great Divide': New social theory on society and nature. Ann Rev Sociol 26:563–584

Gottgens JF (2000) The Paraguay-Paraná Hidrovia: Large-scale Channelization or a 'Tyranny of Small Decisions'. In: Swarts FA (ed) The Pantanal: understanding and preserving the world's largest Wetland. Paragon House, St Paul, MN, pp 135–144

Gottgens JF, Perry JE, Fortney RH, Meyer J, Benedict M, Rood BE (2001) The Paraguay-Paraná Hidrovia: protecting the Pantanal with lessons from the past. Bioscience 51(4):301–308

Habermas J (1987) The theory of communicative action: the critique of functionalist reason, vol 2 (trans. McCarthy T). Polity Press, London

Hamilton SK (1999) Potential Effects of a major navigation project (the Paraguay-Paraná Hidrovia) on inundation in the Pantanal flood plains. Regulated Rivers: Research and Management 15(4):289–299

Hamilton SK (2002) Hydrological controls of ecological structure and function in the Pantanal wetland (Brazil). In: McClain ME (ed) The ecohydrology of South American rivers and wetlands. IAHS special publication no 6, pp 133–158

Ioris AAR (2004) Conflicts and contradictions on the occupation of the Pantanal space. In: Tazik DJ, Ioris AAR, Collinsworth SR (eds) The Pantanal: scientific and institutional challenges in management of a large and complex wetland ecosystem. US Army Corps of Engineers, Washington, DC, pp 26–38

Ioris AAR (2009) Water reforms in Brazil: opportunities and constraints. J Environ Planning Manag 52(6):813–832

Ioris AAR (2010) The political nexus between water and economics in Brazil: a critique of recent policy reforms. Rev Radic Polit Econ 42(2):231–250

Ioris AAR (ed) (2012) Tropical wetland management: the South-American Pantanal and the international experience. Ashgate, Farnham

Ioris AAR (2014) The urban political ecology of post-industrial Scottish towns: examining Greengairs and Ravenscraig. Urban Stud 51(8):1576–1592

Ioris AAR, Irigaray CT, Girard P (2014) Institutional responses to climate change: opportunities and barriers for adaptation in the Pantanal and the upper Paraguay River Basin. Clim Chang 127(1):139–151

Junk WJ, da Cunha CN (2005) Pantanal: a large South American wetland at a crossroads. Ecol Eng 24:391–401

Junk WJ, da Silva CJ, da Cunha CN, Wantzen KM (eds) (2011) The Pantanal: ecology, biodiversity and sustainable management of a large neotropical SEASONAL wetland. Pensoft Publishers, Sofia

Lourival R, de Queiroz Caleman SM, Villar GIM, Ribeiro AR, Elkin C (2008) Getting fourteen for the price of one! Understanding the factors that influence land value and how they affect biodiversity conservation in Central Brazil. Ecol Econ 67:20–31

Meio Ambiente Agora (2011) Operação Guardiões do Pantanal Flagra Crimes Ambientais e Aplica Multas de R$ 122 mil. http://meioambienteagora.com.br/operacao-guardioes-do-pantanal-flagra-crimes-ambientais-e-aplica-multas-de-r-122-mil

Olson DM, Dinerstein E (2002) The global 200: priority Ecoregions for global conservation. Ann Mo Bot Gard 89(2):199–224

PCBAP (1997) Plano de Conservação da Bacia do Alto Paraguai. Brasília, MMA

Piedade MTF, Junk WF, de Sousa Jr, da Cunha CN, Schöngart J, Wittmann F, Candotti E, Girard P (2012) As Áreas Úmidas no Âmbito do Código Florestal Brasileiro. In: Código Florestal e a Ciência. Comitê Brasil em Defesa das Florestas e do Desenvolvimento Sustentável, Brasília, pp 9–17

Pott A, Pott VJ (2004) Features and conservation of the Brazilian Pantanal wetland. Wetlands Ecol Manag 12:547–552

Rabelo MTO, Arts K, Girard P, Ioris AAR, Figueiredo DM (2017) Percepção dos Atores Sociais do Turismo sobre o Pulso de Inundação do Pantanal (MT). Revista Brasileira de Ecoturismo 10(3):708–736

Safford TG (2010) The political–technical divide and collaborative management in Brazil's Taquari Basin. J Environ Dev 19(1):68–90

Sanecap. (2011) Proposta de Projetos: Plano de Metas e Ações de Saneamento Ambiental do Município de Cuiabá. Sanecap, Cuiabá

Schulz C, Ioris AAR, Martin-Ortega J, Glenk K (2015) Prospects for payments for ecosystem services in the Brazilian Pantanal: a scenario analysis. J Environ Dev 24(1):26–53

Schulz C, Martin-Ortega J, Ioris AAR, Glenk K (2017) Applying a 'value landscapes approach' to conflicts in water governance: the case of the Paraguay-Paraná waterway. Ecol Econ 138:47–55

Schulz C, Ioris AAR (2017) The paradox of water abundance in Mato Grosso, Brazil. Sustainability 9(10):1796. https://doi.org/10.3390/su9101796

SEMA (State Environmental Protection Agency-Mato Grosso) (2010) Relatório de Monitoramento da Qualidade da Água da Região Hidrográfica do Paraguai – 2007 a 2009. Cuiabá, SEMA

State of Mato Grosso (2009) Plano Estadual de Recursos Hídricos. Cuiabá, Mato Grosso State Government

Swarts FA (ed) (2000) The Pantanal: understanding and preserving the world's largest wetland. Paragon House, St Paul, MN

TNC-WWF (2011) Ecological risk assessment for the Paraguay River Basin: Argentina, Bolivia, Brazil, and Paraguay. Brasilia, The Nature Conservancy and WWF-Brazil

Tocantins MAC, Sousa WC Jr, Pereira PGP, Guimarães E, Lourival R (2006) Diagnóstico de Política e Economia Ambiental para o Pantanal. Megadiversidade 2(1–2):80–101

Young OR, Berkhout F, Gallopin G, Janssen MA, Ostrom E, van der Leeuw S (2006) The globalization of socio-ecological systems: an agenda for scientific research. Glob Environ Chang 16:304–316

Zeilhofer P, Lima EBNR, Lima GAR (2010) Land use effects on water quality in the urban agglomeration of Cuiabá and Várzea Grande, Mato Grosso State, Central Brazil. Urban Water J 7(3):173–186

Chapter 9
Conclusion: Lessons Learned to Expand Frontier Theory

Abstract Informed by the theoretical and empirical considerations discussed earlier in the book, the final chapter revisits the claim that frontier making has always been fundamental for the circulation and accumulation of capital. The centrality of frontiers is not only due to the demand for minerals, land or other resources, or because frontiers represent fresh market opportunities, but also crucially because it operates as compensation for the saturation of the existing capitalist relations in core areas. At the frontier, the conventional sequence of time and space is suspended and reconfigured, allowing room for the decompression of tensions and contradictions. Consequently, spatial frontiers function as a mirror, where the most bare and explicit features of capitalism are vividly exposed. The final pages examine the meaning and immanence of spatial frontiers, considering them as a laboratory of the historical and geographical agency. It is an invitation for further work and a critical reflection upon the necessity, the configuration and the contestation of spatial frontiers, paying particular attention to the economic and territorial incorporation of the Amazon region and the prospects of political resistance.

Keywords Time-space · Mystification · Risks · Political reaction · Agency

9.1 Spatial Frontiers: Global Frontspaces

Capitalism is often perceived and felt like the elephant in the room. 'It' (i.e. the capitalist relations of production, reproduction and legitimisation) is not a docile Disney-like animal, but moves uncontrollably and causes great damage although only a few people seem to pay serious attention. Those who do notice the beast cannot figure out exactly how to contain it or remove it from the room. And the animal and the people carry on regardless… This crude comparison could go a little further if we imagine an indolent pachyderm wandering around, producing more and more dung, intolerably constrained and longing for more space. (Think about the discomfort of an impounded animal that in its native Africa requires approximately 100 hectares per individual.) The metaphor is illustrative for our purposes here because capitalism

A. A. R. Ioris, *Frontier Making in the Amazon*, Key Challenges in Geography,
https://doi.org/10.1007/978-3-030-38524-8_9

is notoriously harmful, expansionist and strives to take hold of the entire planet.[1] Ever since the golden age of Portuguese navigation, capitalism has systematically and purposefully expanded out of Europe as an unstoppable project with global repercussions. Capitalist activity has taken over continents, locations, cultures and practices throughout the world. In this intense geographical process, frontier making became a central pillar of the circulation and accumulation of capital. Spatial frontiers also function as a mirror, where the bare and explicit features of capitalism are vividly exposed. The most visible features of spatial frontiers are the rapid formation of site-specific mechanisms of resource extraction, economic production and political justification, but the more subtle cultural, social and political dimensions are also of great importance. Because of these interdependent relations between 'core' and 'periphery' through joint processes of exploitation, realignment and reinforcement in both old and new areas, it can be argued that capitalism fundamentally depends on accumulation by frontier making (Ioris 2018a).

Our aim in this final chapter, informed by the theoretical points and the examination of specific locations and sectors in the previous pages, is to examine the meaning and fundamental features of spatial frontiers, considering them as a laboratory of the historical and geographical agency. Bringing together the lessons learned in the preceding case studies, the main purpose here is to suggest a new theoretical approach on frontier and frontier making that is informed but goes beyond the existing accounts. This will involve reflecting on the necessity, the configuration and the contestation of spatial frontiers, beyond the conventional descriptive and quantitative assessments of land use change, resource extraction or commodity production. We also follow the suggestion of Pacheco de Oliveira (2016) that the frontier is a heuristic analytical method. The uneven development of capitalism and the production of frontiers has been the object of a long debate and occupied critical scholars working in several different disciplines for more than a century. The spaces of capitalist frontiers—frontspaces—have been repeatedly studied by scholars, and interpretations range from neoclassical enthusiasm about the economic outcomes of new frontiers to critical voices discussing growing proletarianisation and acts of resistance following the penetration of capitalism. Although often used interchangeably, Watts (2018) differentiates between 'frontier' as a zone of socio-economic advance and 'border' as a line of demarcation between national territories or administrative units. Lund and Rachman (2018) also make a distinction between frontier dynamics (the frontier is a free resource zone, where social order is eliminated, property is disrupted and social contracts dissolved) and territorialisation (when spaces acquire new systems of authority and regulation). Departing from such literature, our focus in these last pages is on the political ontology of frontier making, that is, the locales and landscapes undergoing changes because of the advance of new socio-economic processes that are dialectically connected with wider trends and broader scales.

[1] And even beyond the Earth, as there have been serious attempts to bring capitalist relations to outer space, bypassing the 1967 international treaty, in order to mine asteroids and other planets (The Guardian 2015).

We have been concerned with frontier as an area subject to rapid transformation because of the migration of people and the opening of new economic opportunities, and where authority and governance are significantly diluted and transformed. Even more important is the mediation between central, relatively consolidated areas and the frontier, considering that there are multiple mechanisms of innovation and continuation between 'old' and 'new' areas. Marx argued that capitalism is not merely the movement of exchange values, because circulation alone can never realise capital (considering that the mere exchange of equivalents extinguishes value, and the circulation of money and commodity cannot lead to self-renewal). The success of circulation requires mediation of the total economic process, including geographical connections and socio-ideological aspects, as in the case of new sites for extracting resources and producing commodities. 'Commodities constantly have to be thrown into it anew from the outside, like fuel into a fire' (Marx 1973: 255). New spatial frontiers are therefore required to accommodate economic and social demands and divert attention away from home-grown problems. There is always a demand for materials and resources, new markets and business opportunities, for compensating socioecological degradation and reducing sociopolitical tensions through migration. Marx (1976: 794) observes that some workers emigrate, but 'in fact they are merely following capital, which has itself emigrated'. According to Marx's terminology, a frontier is an area experiencing the development of the conditions and presuppositions for the emergence (and the becoming) of capital. Nonetheless, existing frontier theory clearly needs to be updated and expanded, as proposed below, after some discussion of a highly emblematic example of frontier making.

9.2 Amazon: The Perennial, Textbook Frontier

We will take a final opportunity to make reference to various relevant examples of frontier making, such as Australia and the U.S. and the Midwest, but the South American Amazon will remain our main focus. The Amazon is relevant not only because of the scale and intensity of territorialised change, or its extraordinary culture and biodiversity but because the region has been condemned, first by Europeans and then by national elites, to be a perennial frontier. The Amazon has not become a perpetual frontier because it is huge and remote, rather it has remained distant and alien because it has been perennially misunderstood, devalued and treated as a frontier to be conquered. This is particularly true for the Brazilian section of the Amazon, which comprises around 60% of the vast biome. The whole history of colonisation and nation-building over the last five centuries in the Brazilian Amazon has fundamentally unfolded as the production and reinstatement of frontiers. Bunker (1984) describes the Amazon as an 'extreme periphery' where the economic system is based almost exclusively on the exchange of extracted commodities. Foweraker (1981) defines the Amazon as the 'last great frontier' demonstrating the evolution of the frontier as the gradual advance of capitalist relations and the appropriation of surplus by the rest of the Brazilian national economy. This echoes Heckenberger

(2005), who describes the Amazon as one of the last frontiers of the modern imagination. Even more interesting is the claim by Little (2001) that in the Amazon, frontier making is a perennial phenomenon; the frontier has been opened and closed, and reopened and reclosed, many times, also branching off into numerous subregional frontiers and fractal territories.

It is remarkable how closely the Amazon has been associated with frontier in historical, literary, scientific and official documents. The treatment of the Amazon as a frontier has been a common practice since the early territorial incursions of the sixteenth century, through the harvesting of plant and animal products in the following centuries, and reaching a peak before World War I with the collection of latex (mainly for tyre production). Frontier meant exploration, conquest, migration and dispute with other nations over frontiering projects. It must be noted that the five European powers with stakes in the Americas all maintained territorial interests in the Amazon (Spain, France, Britain, Holland and Portugal, with the last conquering much more territory and establishing a much longer border in the Amazon than the other four nations combined). In 1889 Brazil became a republic, after more than eight decades under Braganza kings, and had to resolve significant border demarcation problems with its Spanish-speaking neighbours. One of the key players in that process was the military engineer, geographer and writer Euclides da Cunha, appointed head of the commission in charge of the Peru–Brazil borders around the Upper Purus River (1904–1905). In his books and unpublished notes, he refers to humans as 'intruders', not really part of a reality that is still 'incomplete' and 'under preparation' (Bolle 2005; Santana 2000). Euclides famously described the Amazon as 'a page of Genesis still to be written' and placed the region 'at the borders of history' (Cunha 2005).

The sensibility of Euclides da Cunha and his attempts to use the cutting-edge geographical and sociological thinking of his time to interpret the remote, forgotten corners of Brazil, suggest what is fundamentally a double estrangement: humans struggling to understand and appreciate the Amazon (obviously Euclides represented the white outsiders and disregarded the knowledge and culture of those groups already living there) and, for its part, the region rejecting preconceived technologies and procedures (because of its lavish ecosystems, large distances, small workforce, tropical diseases and hydrological cycle). This fundamental estrangement, which Euclides recorded during his epic travels, but was also recurrent in the narratives of generations of explorers, is what maintains the Amazon in its condition of the permanent frontier. Such preconceptions and prejudices, which are self-reinforcing, have obviously informed government initiatives and public policy. In the 1930s, the federal Brazilian government launched the 'March towards the West', aiming to explore, occupy and 'fill the gaps' in the national map (gaps which, of course, were already occupied by indigenous tribes, who were never considered or consulted). Between the 1940s and 1960s, successive initiatives and government plans built public infrastructure and encouraged the economic development of the Central-West region (culminating with the shift of the national capital to Brasília) and paved the way for the 'final' subjugation of the Amazon. The process was escalated in the 1970s under the military dictatorship's megalomaniac programme of national integration, fuelled by cheap international credit and Cold War ideology. The process was of course enormously complex and cannot be schematically reduced to simple explanations related to the manipulation of big capital or the protagonism of individuals and peasants (Léna 1986), but requires a proper theory of the frontier, which is discussed below.

The advance of military technocracy and geopolitical fantasy over the Amazon is a well-known story (briefly described in Chap. 3), but something has changed in the last three decades, as regional development has become firmly rooted in the semiotics of free-market ideology and the pursuit of efficiency and market globalisation. In a context of neoliberal reforms (in the 1990s) and neoliberal developmentalism (under the populist administration between 2003 and 2016), large mining, hydropower, timber, road and navigation projects have been implemented. In parallel, socio-environmental legislation expanded, but also became more flexible and increasingly informed by the tenets of ecological modernisation (particularly adopting the language of market solutions and payment for ecosystem services). More areas were opened (deforested) and previous frontiers were renovated to comply with new production procedures. The most extensive and influential case has been the expansion of agribusiness activity to previously forested zones (e.g. Mato Grosso, Rondônia and Amazonas) and former latex production areas (e.g. Pará). Agribusiness, especially soybean production, is now the main mechanism for further embedding the Amazon in the sphere of frontier making. The agribusiness frontier builds on a long trajectory of frontier making, particularly the developmentalist, state-centred process in the 1970s. But the failures and excesses of that nationalistic period (which are generally attributed to inadequate state interventions and a lack of entrepreneurialism, ignoring the actual difficulty of establishing production and trade in the region) are now considered to have been overcome because of the market rationality and productivity of neoliberalised procedures. Agro-neoliberalism is the new frontier of agribusiness in the Amazon, described and praised as rectifying past mistakes and demonstrating the power of plantation farms (Ioris 2017). There have been repeated attempts to contrast the supposed victory of neoliberalised agribusiness—increasingly dominated by transnational corporations and aiming to sell ultra-processed, profit-maximising food—with the previous extractivist basis of the Amazon economy, but the frontier continues to be largely defined by rent-extraction processes and the production of an image of modernity that conceals elements of the colonial past (unrestrained exploitation, racism, slavery, etc.).

The contemporary Brazilian economy depends on the constant movement of agribusiness to new areas, to prevent people from realising that agriculture has been transformed into something beyond food production, among other reasons. Due to the focus on profit and accumulation, agriculture has incorporated ultramodern techniques (digital technology, satellite and Internet-based precision machines, genetically modified crops, powerful pesticides, etc.) and significant amounts of capital (both in direct production and in infrastructure). In other words, the production of new agribusiness frontiers helps to conceal the real results of the meta-agriculture that nowadays produces an increasing proportion of the organic matter consumed by the global population (what looks like food but is really meta-food). Agribusiness frontiers provide legitimacy for the simulacrum of agriculture and food that circulates in global agri-food markets. It is because of frontiers that labour, land and cultures can be subjected to the rationale of neoliberalised agribusiness, given the ideological transfer of responsibility to the private sector (obviously, with the ongoing support and intervention of the state apparatus). Moreover, ultimately the basic elements of frontier making remain the same, only with a new façade dominated by the symbolism of agribusiness. The frontier is now mediated by agribusiness, as well as being the space for the realisation of agribusiness.

Despite the apparent rationality and success of the agribusiness frontier in the Amazon, praised daily by political groups and the mass media, it is in fact the most irrational form of economic development, because of its self-consuming nature. On the one hand, the Brazilian economy is increasingly dependent on the formation of new frontiers. A Financial Times article on 16 January 2018 summarises the macroeconomic role of agribusiness exports, which accounts for 42% of national exports (mainly driven by Chinese demand for soybean). This translates into decisive political power for the agribusiness sector; 44% of the members of the House of Representatives are involved in the industry. Because of their influence, a number of legal adjustments have been introduced or considered in recent years to increase flexibility in the forest code, reduce protection for conservation reserves, adopt ecological modernisation tools (such as payment for ecosystem services), facilitate access for foreign investors and be lenient over land grabbing. Associated with the power of agribusiness, illegal deforestation in the Amazon continues to advance, and currently accounts for 95% of forest loss in Brazil. The perverse logic of Amazonian agribusiness frontiers is also demonstrated by the enormous investment opportunities made available to the economic elites, particularly investment banks, by left-wing politicians who have championed an idiosyncratic form of populist neoliberalism over the last two decades (Ioris 2017). Reflecting upon the centrality of frontier making to capitalism in general, and the perennial status of the Amazon as a triple frontier (of the colonial project, of nation-building and now of globalised markets) in particular, the next section will present elements of a theory of spatial frontiers (frontspaces).

9.3 Theorising Spatial Frontiers: The Time-Spaces of Capitalism

The discussion so far has served to demonstrate that spatial frontiers exist not only because they provide economic and social opportunities that may be more difficult to access elsewhere, but because problems accumulated in central areas are responsible for frontier making (insofar as this constitutes a fundamental endeavour to renovate the whole economic system). At the frontier the new remains tamed, subordinate, relativised as genuine economic creativity because the ontic reality of the frontier is fundamentally shaped by the transplantation of exogenous socio-economic relations. For more than four centuries, there were limited and only sporadic attempts to develop productive activities in the Amazon, due to the more compelling possibility of collecting and exploiting the region's abundant resources, resulting in its perennial frontier status (Bunker 1984). In the middle of the last century, due to the complexification of the Brazilian economy and mounting tensions in the eastern core areas, the Amazon entered a new phase of economic development which, nevertheless, did not change its status as a frontier-maker. On the contrary, the new phase of frontier making was presupposed in the old: the time was ripe for a focus on production (at least the image of production) and for a new language of modernity. Since the 1960s, the region has been a target for the construction of hydropower schemes, roads, mega timber processing plants, navigation infrastructure, the industrial pole of Manaus, and, most importantly, for the advance of agribusiness. Still a frontier, despite looking ever more like a centre (and even more of a frontier precisely because of that).

Consequently, in order to interrogate the frontier, one has to simultaneously comprehend the achievements and contradictions of the most developed (central) economic areas. We can benefit from the methodology adopted by Marx to describe the evolution of capitalism and from his claim that the bourgeois economy supplies the key to the ancient economies that preceded it. Marx (1973: 105) argues that the anatomy of humans 'contains a key to the anatomy of the ape', that is, the intimations of higher development in the less developed systems 'can be understood only after the higher development is already known'. To understand the interpenetration between past and present, Marx specifically warns that one must progress carefully because similarities and differences must be examined in detail, avoiding simplistic, non-critical associations. That is certainly useful, but we need first to reverse the chronological direction of Marx's analysis, from the centre to the (more recent) frontier. The reason is that, although the frontier obviously emerged after the economy evolved in central areas, it nonetheless always revisits the basic elements of the centre's economic past. This is the case regarding, for example, processes of primitive accumulation, dispossession, unregulated appropriation of resources and labour exploitation, ideological argumentation of progress and the imposition of bourgeois order and associated values. Starting from the realisation that frontspaces are predicated and constrained by the lingering failures of the areas where migrants and business people originally came from, the controversial and politicised dynamics of frontspaces can be schematically summarised in five main points that consolidate the findings and provocations of the previous chapters, which are relevant not only for our reference to the Amazon, but for the examination of other contemporary frontiers around the world.

The *first main conceptual claim is that frontier making may appear as a social and spatial dislocation, as distanced from the centre, while in effect the centre is being projected, restated and restored.* There is no essential disconnection between the socio-economy in the consolidated (central) and the new (frontier) areas, but actually a coherent continuity between the centre and the frontier. Frontspaces are therefore more than 'zones of incorporation' of an expanding world system (cf. Wallerstein 1974); they play a key role in the reorganisation and revitalisation of the centre. There is no spatial contradiction between the centre and the new areas, but in reality the frontspace is presupposed in the contradictions of the centre itself and functions as a mechanism to mitigate those tensions and prolong the existence of the centre. In this way, the frontspaces of capitalism are, from the outset, loaded with the practices, realisations and vices of the old areas. For instance, the European colonisation of the Americas transferred to the new continent sociopolitical hierarchies, values and ideologies that underpinned metropolitan societies in Europe. The Americas contained not only riches and resources, but became catalysts for controlled, top-down reforms in the centre. Similar processes were present in the eighteenth century when colonisation expanded in India, and in the nineteenth century during the Scramble for Africa and the partition of the continent among European colonial powers. Simultaneously, other parts of Asia and Oceania were turned into spatial frontiers. In more recent decades several of these former colonies have been recolonised, this time particularly under the rhetoric of market globalisation, foreign investment (often in the form of land, water and nature grabbing) and international development.

In this way, the social, economic and spatial configuration of the frontspace fundamentally replicates the mechanisms of exploitation and exclusion that define capitalist nations (with all their contradictions and frictions). The frontier is more than a spatial fix, as argued by Harvey (2006), that is, a location where capital can be diverted and invested in infrastructure and real estate in order to respond to problems of overaccumlation. Spatial frontiers are areas where time and space are reconfigured and that in theory could result in something new but because of their subordinate status rapidly assume a configuration that largely mimics the core areas. That is, at the frontspace time and space are transfigured but retain the properties they had in the original, central areas. The frontier is where the trajectory of time and space (as established in the central area) is disrupted, creating a realm of potentialities, but then, due to the dynamics of frontier making, the new area is retained within the sphere of influence of the centre. The geography of the frontier fundamentally unfolds around the troublesome gap between the possibility of the new and the concrete reproduction of key features of the (older) centre. As theorised in Aristotle's *Physics*, change requires the existence of potentiality, which is actualised and realised according to specific circumstances but without full independence. The end state of the process of change depends on the specific properties of the system and its potentiality for change. The frontier is less fixed, more tentative and to some extent open, but the previous order there is only superficially and temporarily interrupted, and then rapidly and effectively reconfigured according to what existed before in the centre. The reasons for the stymied potentiality of frontier activity are located in the controlling power of the processes of expropriation, enrichment and authority, which empower some sections of society and establish a new spatial order that only partially, asymmetrically incorporates the majority of the frontier population.

The centre must be upheld (to safeguard dominant interests and political power) through the production of new peripheries, but only to the extent that the periphery mirrors the decisive politico-economic features of the centre. (That is the crucial difference between frontiers in a capitalist economy and refugees from the advance of capitalism, as in the case of alternative communities and isolated indigenous tribes). Although the mindset of frontier people may still not be strictly capitalist (in the sense of efficient realisation of profit and maximisation of economic outputs), the logic of the frontier is indirectly associated with capitalist tendencies operating at different scales. While Harvey (2006) did not pay special attention to the lived processes at the frontiers of capital expansion, his observation that location is an active element in the overall circulation and accumulation of capital is spot on: frontspaces are areas which are still fluid and unsettled, but which from the outset have very favourable conditions for supporting the centre and creating new spaces for capital to flourish. The new spatial frontiers emerge out of the contradictions and constraints of consolidated areas, but also with limited freedom to reconstitute social and economic relations much beyond the given hegemonic conditions. The strategic role of frontspaces is not simply to recirculate capital, but to mitigate mounting social tensions (such as unemployment and political unrest, e.g. Scottish and Irish migrants to British colonies or ex-colonies), release the value of natural resources hitherto beyond economic use (minerals, water, land, labour power, etc.) and pave the way for new cycles of investment and circulation. Frontspaces may be located in either urban or rural areas; in some cases the movement is from conflict-prone rural areas towards the periphery of large cities, as has typically happened in Latin America (the case of Lima's *barriadas* is paradigmatic, exacerbated by Peru's civil war in the 1980s and the liberalising reforms in the 1990s).

Our s*econd main claim is about the production of frontspaces through the reinforcement of the multiple and interconnected dualisms* that characterise Western culture and the Westernised pattern of economic development. Many authors have identified the origins of Western dualisms in Hebrew theology and Greek philosophy, sustained through the Christian polarity of good–evil and the Cartesian mind–body dichotomy that permeate Western patterns of thought and associated scientific knowledge. The result is a situation in which interpersonal dualisms (male/female, white/non-white, learned/traditional, archaic/modern, unproductive/productive, etc.) are deeply rooted in, and help to reinforce, Western culture and its socio-economic structures (Mellor 2000). These ingrained dualisms are forced upon spatial frontiers by dominant interest groups as part of the attempt to consolidate new relations of production and reproduction. The long chain of dualisms derives from the fundamental dichotomy between core and frontier, which is nurtured by the supposed superiority of the centre and the alleged deficiencies of the frontier. The fabricated contrast between 'superior' people in central areas and 'second-class' frontier inhabitants is instrumental both for the institutionalisation of the new frontier and for upholding the authority of the centre. The typical narrative of frontier making replicates the hierarchical differentiation between core and frontier and reinforces the message that activities at the frontier must reproduce social roles and institutions imported from consolidated areas.

Dualistic thinking has been put to work to further the advance and legitimisation of frontier making in different parts of the world. Australian setter culture vividly illustrates the superposition of various dualisms; this society was highly racist, anti-native and hierarchical, permeated by a masculine discourse about the supremacy of the free white settler (Woollacott 2015). Australian frontier making has also been denounced for the recurrent practice of frontier genocide. As described by Rogers and Bain (2016), between 1788 and 1928, extreme brutality was rationalised through the melding of Darwinian ideas about the survival of the fittest (i.e. the white, male settlers) with notions of inferior races that would inevitably die out (i.e. the aborigines). That led to a perverse combination of extinction and extermination, due to the impossibility of pastoralism coexisting with indigenous prairie management. Despite this tragic history, so far there has been limited academic interest in frontier genocide, which Staner (1968, mentioned by Rogers and Bain 2016) calls the 'great Australian silence'. There are still unresolved questions about who should be considered responsible for genocidal colonisation, considering the impact of white settlers, local authorities and, ultimately, the British colonial masters in London. Evans (2007) details the full extent of this genocide, the result of a coordinated onslaught on lives, land and culture, which was central to the evolution of capitalism from the age of mercantilist colonisation to the time of industrialisation. Of course local aborigines tried to resist, sometimes violently, but this 'does not change the fact that genocide occurred' (Rogers and Bain 2016: 90).

The self-professed supremacy of white, Western society and its self-granted permission to conquer and exploit were potentialised during colonisation by strategic scientific developments (navigation, firearms, production tools, etc.) and the application of scientific and religious knowledge (for instance, geography to support imperialist projects, biology and geology to identify valuable resources, and Christian morality to disempower and subjugate the locals). However, old dualisms have continued to reverberate, resulting in accumulated dualities, long after the end of colonialism. The tension between the advance of novel social and economic relations and the persistence of the old values and institutions is one of the main characteristics of frontier making. The range of interconnected dualisms was felt intensely in the Amazon, where the culture, knowledge and skills of traditional peoples, including those who migrated to the region after the 1970s and their families, were systematically devalued to pave the way for the commodification of production and consumption practices. Barbier (2012) asserts that the expansion of the Amazon frontier was in itself dualist, split between agribusiness farms and family agriculture units, while in fact the agricultural frontier is multiple and the different categories of farming are materially and socioculturally interdependent. Nonetheless, the promise of a better life for the large majority of impoverished migrants was never fulfilled in the new reality dominated by large-scale farmers and transnational corporations. The appeal of the modern world, at the expense of social traditions and community life, is also illustrated by the ongoing advance of processed, frozen food into the most remote corners of the region (with all the associated problems for health and the local economy), as in the case of the upper Negro River Basin on the border between Brazil and Colombia.

The *third ontic feature of spatial frontiers is dialectically related to the previous two: time and space at the frontier are compressed, reconfigured and launched in different directions*. Spatial and temporal changes do not necessarily progress in the linear and sequential manner typical of core areas; at the frontier the basic mechanisms of expropriation, commodification, proleterianisation, etc. will follow unique patterns (obviously connected with the wider socio-economic trends and structures).

The frontier has different phases, which normally begin slowly and then, when circumstances are favourable, accelerate rapidly. The frontier's very existence is never guaranteed, but one frontier can open and close several times on different occasions. A particular area that was considered a functional frontier for the purpose of capitalist relations can suddenly lose that status, for instance, due to competition from new products or production areas. Then, after some time, what had become an obsolete frontier can be recreated and incorporated into new rounds of migration and production as new opportunities and additional technologies become available (e.g. the handling of new products and goods in spaces previously used for others). This means that old frontiers are excavated through the redeployment of knowledge and practices that, once again, are externally imposed from the centre—if the frontier could re-emerge independently of the centre it would no longer be a frontier. Not only can old frontiers be supplanted by new ones, but different 'frontier moments' can be both imposed and superposed on previous socio-spatial experiences. Ultimately, 'frontier is not space itself. It is something that happens *in* and *to* space. Frontiers *take* pace. Literally' (Rasmussen and Lund, 2018: 388).

New socio-spatial relations are built upon past experiences, not necessarily improving practices or procedures; in fact, a spatial frontier may be new in historical terms while still replicating some of the oldest and vilest relations and institutions. At the frontier, capitalism is not only renewed through novel technologies and productive platforms but is also virtually free to reinstate elements of violence, exploitation, dispossession, racism and other injustices that characterised previous historical periods. This has been the case with the Amazon region, which was a frontier of biodiversity and mineral extraction during colonial times, then the main source of plant latex at the time of the Second Industrial Revolution and, more recently, due to the demand for agribusiness goods, has become a dynamic frontier for plantation production and export. Violence was employed as a central element of colonisation strategy, and the expanding frontiers pushed forward by the invading Europeans did not in fact advance civilisation but rather destroyed social groups and their sophisticated knowledge and art (Hemming 1987). One of the most notorious examples was the legislation introduced by the Marquis of Pombal, prime minister of Portugal, in 1757 (called *Diretório*) which forced indigenous groups in the Amazon to move to settlements managed by a 'director' [*diretor*] where racial assimilation was encouraged and cultural and linguistic identity subsumed. During the rubber extraction period in the nineteenth century, the existence of indigenous groups was ignored and contingents of very poor migrants were attracted to the region to collect latex, which in the end served to enrich a very small elite in Manaus and Belém while satisfying the growing industrial demand for natural rubber. This exemplifies how genocide, slavery and violence were not sporadic incidents, but constituted an ongoing, systematic and transnational phenomenon underpinning frontier making.

The *fourth element of our conceptualisation is the mystification of the benefits and opportunities available at the frontier*. Any frontspace is always highly hierarchical and often manipulated to serve mainly the interests of those in more favourable positions (which include land speculators, rural development companies, intermediaries and fixers, opportunistic investors, and traders who receive and export goods produced by a large number of individual agents), but these asymmetries are disguised by the appearance of accessibility and better prospects for earning a living. The mystification of what fronstpaces are really offering is based on deliberate misrepresentations or omissions. There exists a fetishism of the frontier that is nourished by ambiguous evidence of success and vague stories about people who thrived. Turner (1920) misrepresented the frontier as a conduit of democracy and equality, and his account exemplifies the positive narrative constructed by those who gain from the frontier. Along similar lines, Bowman (1927: 64) argued in his work on 'pioneer fringes' that a 'changing environment breeds liberalism if the resources are abundant enough to support close settlements and the development of independent social and political institutions.' However, instead of a political vacuum, the frontier is a space of social control where autonomy was a clear strategy for governing the territory (Hogan 1985). The mystification of the frontier also determines that success is measured according to the values of central areas and Western standards. Failures are seldom attributed to frontier conditions; rather, the blame is placed on the incompetence of migrants and pioneers who failed to take advantage of the opportunities presented to them.

The memories of those who migrated from the south of Brazil to the Amazon were populated by images of courage associated with their Italian and German ancestors (who moved to South America in the nineteenth century) and the mythology of bravery related to the consolidation of international borders with the Spaniards, and later the Argentineans. Wealthy landowners, subsistence farmers and workers were all products of the same agrarian past lived in the south of the country, but the symbolism of the frontier was appropriated differently by different social groups. The rhetoric of victory and anticipated success was repeatedly invoked by those who led the opening of new agricultural areas in the Amazon in the 1970s, which helped to downplay the obstacles faced by the newcomers and the crude reality of socioecological exploitation. The plan of the military dictatorship was to allocate land to impoverished peasants and landless people (considered troublemakers in their areas of origin), but it was a monumental disappointment as it largely failed to foster agrarian capitalism in the region (Rivière d'Arc and Apestéguy 1978). The majority of those who migrated to the Amazon did not have the means to secure or maintain land and ended up as proletarians in urban or rural areas. In practice, the frontier was less epic and more a daily fight for survival (a significant proportion of migrants did not find success and returned, even more impoverished, to their areas of origin in the south, where many took part in other forms of protest and land occupation).

The Australian frontier-making experience, which has significant differences from that of the Amazon and other parts of South America, is frequently associated with an image of conquest and triumph, but this is largely explained by the construction of a new society that mirrored British values and social hierarchies. Australia in the early nienteenth century was a major destination for convicts and ex-convicts (whom the authorities wished to remove from the motherland) and then, after the 1840s, a target for free settlers, encouraged to move to the colony to take control of 'free land', in fact aboriginal land (Woollacott 2015). As far back as 1834, an

Australian settler argued that those who colonise a new area 'are sure to enjoy a greater degree of consideration and importance among their companions than they could reasonably have hopped to attain in the older society' (Wakefield 1967: 327). Inequalities were not only found between settlers and aborigines; a small squatter elite (described as a 'squattocracy', and including, among others, members of the Melbourne Club, established in 1839) controlled most of the land and limited access to new migrant contingents. This created serious resentments and pressure for land reform. New legislation was introduced, such as the 1860 Nicholson Land Act, but its effectiveness was limited, as the powerful squatter elite could still purchase whatever land they required through the use of 'dummy bidders'. Wealthy squatters also used their knowledge of the land to buy up the best locations, leaving only the least fertile ground for less-privileged settlers.

The mystification of frontier opportunities was also an important feature of the conquest of the American Midwest around Chicago. After 1833, the regional population and the economy grew rapidly due to the activity of settlers, investors and speculators. These so-called 'boosters' advanced economic theories and well-crafted rhetoric about the natural endowments of the Chicago area (minimising the obvious need to invest in infrastructure and logistics to make the frontier really flourish). Their actions were typically loaded with enthusiastic exaggeration and self-interested promotion. Due to improvements in rail and boat transportation, their promotion of the frontier became a self-fulfilling prophecy and during the nineteenth century Chicago became the main trade centre for grain, timber and meat (Cronon 1992). The accomplishments of the frontier were internationally celebrated, culminating with the staging of the 1893 World's Columbian Exposition in Chicago to applaud the progress of American civilisation and the fulfilment of Columbus' dreams four centuries earlier (obviously discounting the immense social and environmental impacts). It is interesting that, while mystification and ideological pressures were extremely powerful, these phenomena also flourished because people in their daily struggle for survival are often led to express conservative views and put up with the current state of affairs. The most vulnerable and disorganised groups at the frontier tend to have serious difficulty developing coordinated opposition, despite their actual level of consciousness.[2]

Fifth, even in a globalised and highly interconnected world frontier dynamics will not die out. On the contrary, *frontier making will continue to expand and flourish around the world either through the incorporation of new areas hitherto subject to less capitalist influence, or with the replacement of previous frontier-making activity with novel rounds of capitalist relations of production and reproduction.* Accelerated market fluxes and population mobility do not dispense with spatial frontiers, because the modern Western world persistently strives for new places and landscapes to conquer (Ioris 2018b). One of the decisive features of capitalist modernity is how

[2]In his study on territorial conquest and European border disputes, Namier (1942: 69–70) perspicaciously observed that: "One would expect people to remember the past and to imagine the future. But in fact, when discussing or writing about history, they imagine it in terms of their own experience, and when trying to gauge the future they cite supposed analogies from the past: till, by a double process of repetition, they imagine the past and remember the future".

it aims to standardise location-specific processes and incorporate them into the same market-centred rationality. Frontspaces are fertile areas for the reaffirmation of Western modernity because existing socioecological elements of their reality are typically disorganised and thus easily replaced with new features associated with the Global North. However, it is important to realise also that capitalism does not need frontiers merely to renew itself, but, on the contrary, frontier making helps the centre to remain largely as it has always been (see above). At the spatial frontiers, capitalism can be more capitalistic, in the sense that it is less constrained and more potentialised by the unique conditions of frontspaces.

There exists a necessity for new spatial frontiers to work as opportunities to try to erase socioecologies and produce novel spaces or new socio-economic relations under the influence of capitalist modernity. Frontier making is intrinsic in the peculiar trajectory of capitalism that combines dualisms and accumulates tensions between old and new, exploitation and production, particularities and universalisms. But the necessity for spatial frontiers needs to be understood in dialectical and non-prescriptive terms. Žižek (2011) claims that historical necessity is really a convergence of contingencies. He argues that the Hegelian notions of totality and historical necessity are in fact elements of flexible reasoning that imply a radically open contingency of history. The relation between contingency and necessity is dialectical, in the sense that there is a necessity for contingencies and, more radically, a contingency of necessities (i.e. things became necessary only in a contingent way). The relation between past and present is also dialectical, as the present is obviously influenced by the past, but the past is also reinterpreted and reconstructed by the present. As also observed by Bukharin (1929), necessity is really a chain of historical events that connect cause and effect. Rather than the trends of history being determined a priori by some overpowering force, historical necessities can only be explained retrospectively. In this way, Hegelian necessity should be seen not as a cause, but as the central property of the process of change (Mann 2008). The notions of historical necessity and dialectics are particularly relevant to understanding the Hegelian theorisation of global trends and the interventions of the state apparatus (Ioris 2014), as much as to the search for alternatives to capitalist frontiers, as examined next.

9.4 Immanence and Interstices: Agencies in the Unexpected

The five main ontic features of frontspace discussed above constitute an attempt to identify, if only schematically, the basis for the insertion of new areas and reinsertion of old ones into the sphere of influence of Western, capitalist modernity. Spatial frontiers continue to expand, including processes of production, extraction and politico-ideological containment, not only because of favourable economic opportunities but mainly due to the need to stabilise and invigorate core economic and political trends. At the frontspace, capitalism can reassert its hegemony with much lower costs and fewer restrictions. At the frontier, order, authority and convention are suspended, time and space acquire new meanings, excesses are committed—Martins (2009: 09) describes the frontier expanding into the Amazon as the ‘scenery of intolerance, ambition and death'—but the compelling symbolism of abundance, potential wealth and a bright future represents a powerful legitimisation tool. This process has not been interrupted by market-based globalisation, because frontier making continues to be predicated in rising tensions in central economic areas and their unstoppable demand for goods and services (Ioris and Ioris 2018). Such inexorable interdependency between centre and frontier is clearly a process with major ideological and political significance. The evolution of the frontier could hypothetically take any direction and lead to different social arrangements, but in practice there is a great deal of constraints due the hegemony of capitalist relations. It is a self-fulfilling prophecy normally recounted only by the winners. The success of frontier making, from the perspective of Western economists and policymakers, depends on the consolidation of private property, the ability to exploit socio-nature, incentives for the circulation of capital and widespread commodification and financialisation. Experiences that deviate from this model are considered anomalies and curiosities rather than genuine, viable frontiers.

The organisation of capitalist frontiers affects the maintenance of what Rancière describes as the police order, a symbolic constitution of the social that both fragments and incorporates, insists on homogenisation and pushes for consensus. The individual must passively comply, circulate in a space emptied of politics, as 'the space of circulation is nothing other than the space of circulation' [*l'espace de la circulation n'est que l'espace de la circulation*] (Rancière 1998: 242). In agribusiness frontiers such as the Amazon, crop monoculture has disturbing material and symbolic parallels with political and cultural monocultures that permeate highly hierarchical and exclusionary societies. The result is a consensus on the inegalitarian order [*consensus sur l'ordre inégalitaire*] (Rancière 1998: 74), which can and must be disrupted with the emergence of politics (which for Rancière means, fundamentally, emancipation and radical equality). If the frontier space is unique in terms of potentiality and transformation being contained by the hegemony of capitalist prerogatives, these mechanisms of control and legitimisation are not absolute. On the contrary, because of the fluid boundaries and high mobility of newcomers, the frontier contains interstices in which political consciousness and reactions can emerge. Resistance to the capitalist logic of the frontier is not something that flares up by chance, but is located in the very constitution of the frontier. The frontier has multiple time-space discontinuities as much as it shows rugosity (which are leftover characteristics from prior periods, according to Santos 1985) and it is in this context that opposition, almost always silent, but sometimes intense, erupts.

The frontspace entails dialectics of apparent potentiality and concrete results that are not far removed from conditions in the centre. Nonetheless, this does not mean that historical and geographical agency is stifled; on the contrary, frontspaces are experienced as imagined, lived and disputed landscapes. It is through the cultural and physical working on the landscape that the need for frontiers can be rejected and gradually subverted. There is historical and geographical necessity that comes from the mounting contradictions of the frontier. The main driver is not people's pre-established category or affiliation, but the political subjectivity that allows them to become effective agents of transformation. This form of politics is more than mere opposition between oppressors and oppressed, or between the police order (the existing government, judicial and parliamentarian institutions) and politics (the rupture produced by people and their demands for equality), but it is intrinsically impure, in the sense that it is multiple and multiplied (Chambers 2011). According to Rancière (1998), the political establishment is unable to include all social segments and fairly distribute the shares of economic activity and social achievements; the logic of the police is to manage inequality and secure legitimacy, denying the existence of different needs and social difference. But reactions to the political order come when something is perceived to be fundamentally 'wrong' [*le tort*] which increases the chance that people will become politically engaged.

Resistance is in the translocation of economic and social patterns from the centre to the frontier while both migrants and established residents retain complex memories and experiences from pre-frontier times that allow them, depending on circumstances, the possibility to imagine some form of alternative. Durkheim (*On Suicide*) demonstrated that the most personal problems have sociological bases and spatial expressions; in our case, the frontier is a sociological process subject to significant individual forces. Resistance and reactions are informed and fuelled by past individual and collective experiences, such as previous conflicts or repression elsewhere. Depending on how people became involved and were relocated to the frontier (spontaneously or via government agencies), they will be more or less willing to question authority and risk whatever they have. The past is mobilised and influences the present because of cultural proclivities and subjective attitudes related to particular experiences which ensure that particular kinds of historical consciousness become meaningful (Whitehead 2003). The frontier is a cross-scale process that results from national and international pressures, but the interplay between culture and history (beyond any false dichotomy between these) is resolved at the level of landscape change. The crafting of landscapes encapsulates historical and political consciousness that help to shape group identity. The landscape of large agribusiness areas is the consequence of multiple agencies that both converge and diverge, as the heavy machinery of wealthy landowners contrasts with pockets of family agriculture and those living along roads and in marginalised areas. That is related to what Quaini (1982: 171) describes as the 'geo-history of power'.

Labour is culturally situated, and culture has political consequences, as 'the forms of unity and opposition take on the *cultural* expression they do in specific political struggles' (Smith 1989:15). It is through labour (and constant threats to livelihood and survival) that culture and class-based politics are connected. Particularly because of the heterogeneity of the frontier, labour relations involve not only conflict but also interdependencies between wealthy players and the majority of society. The cultural construction of the frontier as a space of opportunity and likely rewards for those who persevere helps to maintain social inequalities, under the assumption that social mobility merely depends on hard work (and a bit of luck). At the same time, resistance can flourish in the interstices of capitalism's frontier project. Although open rebellion and confrontation are rare, the main form of resistance is a silent process that happens through various forms of practices and positions (Scott 1985). For instance, the main economic activity in the Amazon today is agribusiness, which

Fig. 9.1 *O Catador de Soja* (The Soybean Collector), by Wander Melo

is increasingly financialised (hence more abstract than ever) and unable to address the mounting crises of its own making, meaning that it is increasingly possible that hegemonic trends will be challenged.

Figure 9.1 illustrates the multiple agencies inscribed in the landscape of the agribusiness frontier in our main case study areas in Mato Grosso. It is possible to contrast the heavily modified space of the grain processing plant, with its warehouses, planted trees and sentry boxes, with the background hills of the space that 'was there before'. The main focus of the painting is the most forgotten, least important character in the whole soybean industry: the *catador de soja* (soybean collector), someone who survives by cleaning the floor of the lorries that have recently discharged their cargo. Normally men, these individuals spend most of the day collecting discarded grains that have become only waste. These are marginalised people who work in the lowest fringes of the regional economy, filling a bag with grain and selling it for some food at the end of a long journey. The condition of destitute persons in the north of England collecting pieces of coal from the mining operation to mitigate unemployment and to obtain some essential domestic fuel was famously described by George Orwell. 'All day long over those strange grey mountains you see people wondering to and fro with sacks and baskets among the sulphurous smoke (many slag-heaps are on fire under the surface), prising out the tiny nuggets of coal which are buried here and there. (...) In Wigan the competition among unemployed people for the waste coal has become so fierce that it has led to an extraordinary custom called 'scrambling for the coal', which is well worth seeing. (...) Technically it is stealing but, as everybody knows, if the coal were not stolen it would simply be wasted' (Orwell 1989: 93–95). The beauty of the Brazilian painting, in addition to its aesthetic value, is in how it subverts the established order of things (the police order) and brings to the forefront of history someone who is considered a beggar, a pariah and worthless. The artist, Wander Melo, makes a very compelling point and the painting is certainly thought-provoking: who is really holding historical agency here? The lorry drivers, the staff of the grain corporations (who barely appear in the image) or the *catador*, the poorest of the poor, but the only one, along with his dog, who actually has autonomy? The image is a wonderful representation of the interstices of the oppressive agribusiness activity, an unexpected moment when the subaltern becomes the protagonist, when the wonders of capitalist modernity are minimised and the underdog (with incidentally another underdog) takes centre stage.

Ultimately, the prospects of consciousness and resistance based on daily survival through the production of new landscapes have major implications for a critical research agenda. Interrogating the frontier is a formidable challenge for critical, left-wing thinking (primarily concentrated on justice and equality), considering that frontier making is by definition a generation and perpetuation of inequalities. Likewise, critical scholars need to develop the ability to work through the political, apparently chaotic process of landscape change and silent resistance through the interstices of the established, taken-for-granted foundations of the frontier. All this requires a serious reflexive commitment and rejection of positivistic, politically void accounts of frontier making. As advocated by Lacoste (1973, in Quaini 1982), we need to reflect in order to measure, and not measure first to reflect later. That does not mean looking for facts to fit the conceptual model, but rather a firm investigative effort that amalgamates comprehensive empirical data and constant critical thinking. To a certain extent, this line of investigation will pick up the analysis where Marx left off, in the sense that the richness of his investigation, particularly in his final years, was disregarded and did not survive into the twentieth century (Jones 2016). Marx struggled producing a general theory of capitalism and explain the so-called 'expanded reproduction' (which was only sketched and left incomplete in *Capital*). At the same time, he increased his geographical sensibility and acknowledged the importance of local circumstances beyond Western Europe, the strategic role of the state and the significance of rural communities and common properties (as in the case of Russia). This opens the door to embrace cultural and institutional complexities and reject linear, preconceived trajectories of change (as exemplified by the dogmatism of Leninist-Stalinist formulations). Frontier may be a tentative, uncertain space where capitalist relations more easily prevail, but it can also be seen as a frontier of resistance and of overcoming the perverse socio-spatial consequences of capital's ascendancy.

References

Barbier EB (2012) Scarcity, frontiers and development. Geogr J 178(2):110–122

Bolle W (2005) O Mediterrâneo da América Latina: A Amazônia na Visão de Euclides da, Cunha. Revista USP 66:140–155

Bowman I (1927) The pioneer fringe. Foreign Aff 6(1):49–66

Bukharin N (1929[1917]) Imperialism and world economy. Martin Lawrence, London

Bunker SG (1984) Modes of extraction, unequal exchange, and the progressive underdevelopment of an extreme periphery: the Brazilian Amazon, 1600–1980. Am J Soc 89(5):1017–1064

Chambers SA (2011) Jacques Rancière and the Problem of Pure Politics. Eur J Polit Theory 10(3):303–326

Cronon W (1992) Nature's metropolis: Chicago and the Great West. Norton and Company, New York

Cunha ERP (2005[1909]) À Margem da História. Academia Brasileira de Letras, Rio de Janeiro

Evans R (2007) A history of Queensland. Cambridge University Press, Cambridge

Financial Times (2018) Brazil: rainforest pays the price for the country's crisis, 16 Jan 2018. https://www.ft.com/content/971f03aa-f9e0-11e7-9b32-d7d59aace167

Foweraker J (1981) The struggle for land. Cambridge University Press, Cambridge

Harvey D (2006[1982]) The limits to capital, new edn. Verso, London and New York

Heckenberger M (2005) The ecology of power. Routledge, Abingdon

Hemming J (1987) Amazon frontier: the defeat of the Brazilian Indians. Papermac, London

Hogan R (1985) The frontier as social control. Theory Soc 14(1):35–51

Ioris AAR (2014) The political ecology of the state. Routledge, London

Ioris AAR (2017) Places of agribusiness: displacement, replacement, and misplacement in Mato Grosso, Brazil. Geogr Rev 107(3):452–475

Ioris AAR (2018a) Amazon's dead ends: frontier-making the centre. Polit Geogr 65:98–106

Ioris AAR (2018b) Place-making at the frontier of Brazilian agribusiness. GeoJournal 83:61–72

Ioris RR, Ioris AAR (2018) Colombia's fractured history and continued challenges following the Havana accord. J Peace Build Dev 13(1):79–83

Jones GS (2016) Karl Marx: greatness and illusion. Penguin, London

Léna P (1986) Aspects the la Frontière Amazonienne. Cahiers des Sciences Humaines 22(3–4):319–343

Little P (2001) Amazonia. Baltimore, John Hopkins

Lund C, Rachman NF (2018) Indirect recognition. Frontiers and territorialization around Mount Halimun-Salak National Park, Indonesia. World Dev 101:417–428

Mann G (2008) A negative geography of necessity. Antipode 40(5):921–934

Martins JS (2009) Fronteira: A Degradação do Outro nos Confins do Humano. São Paulo, Contexto

Marx K (1973) Grundrisse (trans: Nicolaus M). Penguin, London

Marx K (1976[1867]). Capital, vol 1 (trans: Fowkes B). Penguin, London

Mellor M (2000) Feminism and environmental ethics: a materialist perspective. Ethics Environ 5(1):107–123

Namier LB (1942) Conflicts. Books for Libraries Press, New York

Orwell G (1989[1937]) The road to Wigan Pier. Penguin, London

Pacheco de Oliveira J (2016) O Nascimento do Brasil e Outros Ensaios. Contra Capa, Rio de Janeiro

Quaini M (1982[1974]) Geography and Marxism (trans: Braley A). Basil Blackwell, Oxford

Rasmussen MB, Lund C (2018) Reconfiguring frontier spaces: the territorialization of resource control. World Dev 101:388–399

Rancière J (1998) Aux Bords du Politique. Gallimard, Paris

Rivière d'Arc H, Apestéguy C (1978) Les Nouvelles Franges Pionnières en Amazonie Brésilienne. Études Rurales 69:81–100

Rogers TJ, Bain S (2016) Genocide and frontier violence in Australia. J Genocide Res 18(1):83–100

Santana JCB (2000) Euclides da Cunha e a Amazônia: Visão Mediada pela Ciência. História, Ciências, Saúde – Manguinhos 6(suppl):901–917

Santos M (1985) Espaço e Método. São Paulo, Nobel
Smith G (1989) Livelihood and resistance. University of California Press, Berkley
Scott JC (1985) Weapons of the weak: everyday forms of Peasant resistance. Yale University Press, New Heaven and London
The Guardian (2015) Silicon Valley exploits time and space to extend the frontiers of capitalism, 29 Nov 2015. https://www.theguardian.com/commentisfree/2015/nov/29/silicon-valley-exploits-space-evgeny-morozov
Turner FJ (1920) The frontier in American history. H. Holt and Company, New York
Wallerstein I (1974) The modern world-system I. Academic Press, New York
Watts MJ (2018) Frontiers: authority, precarity, and insurgency at the edge of the state. World Dev 101:477–488
Whitehead N (ed) (2003) Histories and historicities in the Amazon. University of Nebraska Press, Lincoln and London
Wakefield EG (1967[1834]) England and America. Augustus M. Kelley, New York
Woollacott A (2015) Settler society in the Australian colonies. Oxford University Press, Oxford
Žižek S (2011) Hegel's century. In: Žižek S, Crockett C, Davis C (eds) Hegel and the infinite. Columbia University Press, New York, pp ix–xi

MIX
Papier aus verantwortungsvollen Quellen
Paper from responsible sources
FSC® C105338

If you have any concerns about our products,
you can contact us on
ProductSafety@springernature.com

In case Publisher is established outside the EU,
the EU authorized representative is:
Springer Nature Customer Service Center GmbH
Europaplatz 3, 69115 Heidelberg, Germany

Printed by Libri Plureos GmbH
in Hamburg, Germany